Logic PhDs
Volume 1

Foundations of Combinatory Logic
Grundlagen der kombinatorischen Logik

Volume 1
Foundations of Combinatory Logic: Grundlagen de kombinatorischen Logik. H. B. Curry, translated and presented by Fairouz Kamareddine and Jonathan Seldin

Series Editor

Jean-Yves Beziau   jyb@jyb-logic.org

Foundations of Combinatory Logic
Grundlagen der kombinatorischen Logik

Haskell Curry

Translated and presented by
Fairouz Kamareddine
and
Jonathan Seldin

© Author, translators and College Publications 2016.

All rights reserved.

ISBN 978-1-84890-202-2

Curry, H. B.. "Grundlagen der Kombinatorischen Logik." *American Journal of Math* 52:3, 509-536 and 52:4, 789-834, 1930.
© 1930 The Johns Hopkins Press. Translated and reprinted with the permission of Johns Hopkins University Press.

Published by College Publications

http://www.collegepublications.co.uk

Cover design by Laraine Welch
Printed by Lightning Source, Milton Keynes, UK

All rights reserved. No part of this publication may be reproduced, stored in a retrieval system or transmitted in any form, or by any means, electronic, mechanical, photocopying, recording or otherwise without prior permission, in writing, from the publisher.

The Work of H. B. Curry on Combinatory Logic

FAIROUZ KAMAREDDINE AND JONATHAN P. SELDIN

A. Circumstances of Curry's PhD and its Content

This dissertation, *Introduction to Combinatory Logic*, represents a first major work on a new subject, on which there were only two previous publications, [50] and [14]. Of these, [50] introduced the idea for the system but presented no basic axioms and theorems, and [14] presented the formal foundations of the system and proved a few preliminary results, but was very preliminary. The dissertation, which gave the name *combinatory logic* to the system, was the first publication to give a complete formal introduction and formal development. It was written at Göttingen, nominally under the direction of David Hilbert, although Curry actually worked mostly with Paul Bernays.

The dissertation gives a full definition of combinatory logic as a formal system in which the terms are built up from atomic terms, which consist of *variables* and the constants B, C, K, W, Q, Π, P, and Λ by means of the operation, called *application* of forming (XY) from X and Y. Here, the terms are thought of as functions and (XY) represents the operation of applying the function X to the argument Y; here any term can be applied to any other term, and the result of applying an argument to a function may be another function. Thus, what in ordinary mathematics is written as '$f(x)$' is written '(fx)', and it is assumed that $fxyz$ means $(((fx)y)z)$. The terms B, C, K, and W are called *basic combinators* and are defined by the equations

$$Bxyz = x(yz),$$
$$Cxyz = xzy,$$
$$Kxy = x,$$
$$Wxy = xyy.$$

In [14], Curry followed Schönfinkel [50] in using as the basic combinators K and S, where S is defined by the equation

$$Sxyz = xz(yz).$$

At this time, Curry did not understand S the way he understood the others. This S can be defined in terms of B, C, K, and W, as can the

combinator I be defined by the equation

$$Ix = x.$$

Curry defined equality by means of a term Q so that '$x = y$' is an abbreviation for 'Qxy'. His axiom for equality was WQX, which implies QXX, which is abbreviated as $X = X$. There were also *combinatory axioms*. The rules included rules implying the above axioms defining the basic combinators, rules for Q, including Leibniz' rule, what we now call the elimination rules for Π (universal quantifier), P (implication), and what we now call the introduction rule for Λ (conjunction).

The *combinatory axioms*, taken together, are equivalent to adding to the rules and the axiom for Q the rule that if $x_1, x_2, \cdots, x_n$ are distinct variables none of which appear in U and V, and if $Ux_1x_2\cdots x_n = Vx_1x_2\cdots x_n$, then $U = V$. This extensionality principle made Q equivalent to what we now call combinatory $\beta\eta$-conversion.

After proving the basic properties of the combinators themselves, he took up the proof of the consistency of the system. His major difficulty was that the only reduction relation he had was what we now call weak reduction. This made it impossible to prove the Church-Rosser Theorem, that if two terms are equal, there is a term to which both reduce, which is now the standard way one proves the consistency of systems of this kind. Instead, he was looking at sequences based on what we now call weak contractions with strings of variables added on the right. He was able to prove that if $\mathfrak{X}$ is any term of the system in which the free variables $x_1, x_2, \cdots, x_n$ may appear, then there is a term X in which none of these variables appear with the property that

$$Xx_1x_2\cdots x_n = \mathfrak{X}.$$

It follows from the extensionality principle that any two such Xs for a given $\mathfrak{X}$ are provably equal. Curry also proved the consistency of the system.

B. Background in terms of the history of logic and of Curry's experience

Curry had become interested in logic while a graduate student at Harvard. When he was only 21 years old, he looked at the first chapter of [64], and noticed that of the two rules of inference presented there, one, the rule of substitution of well-formed formulas for propositional variables, was considerably more complicated than the other, the rule of detachment (which is equivalent to modus ponens). This is true even though the substitution involved does not involve the complication of bound variables. (It has turned out that most, if not all, of the attempts to correctly define substitution with bound variables before the publication of [36] were in error.)

The complication Curry noticed in the rule of substitution in Chapter 1 of [64] is now considered to be the complication of its implementation by a computer program, although there were no electronic computers when Curry noticed this (in 1922). He set about trying to find a simpler form (or forms) of this rule, and this led him by about 1926 to some of the combinators. These are largely the combinators I and those to which Curry gave names different from those Schönfinkel had given them.

Once Curry had this idea, it was clear that there was nobody at Harvard who could supervise a dissertation on this subject, so for the year 1927–28, he obtained a one-year position as an instructor at Princeton. During that year, as a result of a library search, he discovered the paper [50], which shocked him, since he had thought his ideas were completely original. Curry immediately ran to see Oswald Veblen, who calmed him down by saying to him, "Good. I like it when somebody else has had the same idea I had because it shows that I am on the right track." Veblen then helped Curry try to find out more about Schönfinkel. It happened that the topologist P. S. Alexandroff was at Princeton at that time, and from him both Veblen and Curry learned that Schönfinkel was in a psychiatric facility in the USSR and would probably not do any more mathematical work. The paper [50] was a report of a talk Schönfinkel had given at Göttingen in 1920, and that paper had actually been written up for publication by Heinrich Behmann. It thus appeared that for Curry to write a dissertation on this subject, he should go to Göttingen, where there were people who knew about the subject. Curry wrote [14] as part of an application for a grant to go there. He did not get that grant, but he got another one, and so he was able to go there in the fall of 1928 and write the dissertation.

C. Further Developments

After finishing the dissertation, Curry obtained a position on the faculty of the Pennsylvania State College at State College, Pennsylvania, and started teaching there in September, 1929. He felt isolated there because he had been at both Harvard and Princeton, but there was no other possibility of an academic position for him at that time. So he started teaching there and also began a series of papers extending combinatory logic, which he hoped would lead to a system that could include all of mathematics. Curry did obtain a National Research Fellowship in 1931, which was originally supposed to last two years, but after the first year his second year was cancelled on the grounds that he had a job to go back to while other National Research Fellows did not. (This was, after all, near the bottom of the Great Depression.) Curry had spent that first year at the University of Chicago, and had thoughts of spending the

second year in Europe, but that was not to be.

In one of the papers Curry wrote during this period, [16], Curry defined

$$[x_1, x_2, \cdots, x_n]\mathfrak{X}$$

to be one of the terms X in which none of $x_1, x_2, \cdots, x_n$ occurs for which

$$Xx_1x_2\cdots x_n = \mathfrak{X}.$$

(By the existentionality property that follows from the combinatory axioms, all such terms X can be proved equal.) This is now called the *bracket abstract*.

In other papers in this series, Curry had dealt with the universal quantifier [15], with implication and equality [17], and with what he called *functionality* [18], which became the basis of what we now call type assignment. Curry called systems of this kind *illative combinatory logic*. Eventually, by the 1960s, illative combinatory logic came to include any system of combinatory logic with atomic constants that were not basic combinators, and so illative combinatory came to be equivalent to applied λ-calculi.

In 1933, Curry learned of [7], which defined a formal system based on what is now called λ-calculus. At about this time, one of Church's graduate students, J. B. Rosser, wrote to Curry that he was writing at Church's suggestion about the work he and S. C. Kleene were doing under Church's direction, since it seemed to be related to what Curry was doing, and Curry, Kleene, and Rosser were friends from then on. During the following year, 1934, Kleene and Rosser proved that Church's system of [7] and Curry's system from [17] are inconsistent; see [42]. They did this by deriving a form of Richard's Paradox in each of the systems. Church and his students reacted to this by extracting the λ-calculus from the system of Church and abandoning the idea of using it as a basis for a formal system to include all of mathematics.

The λ-calculus is now considered to be a formalism essentially equivalent to combinatory logic. The pure λ-calculus consists of terms made up from variables by application and also by *abstraction*, which for any term M and any variable x forms the *abstraction term* $\lambda x . M$. This abstraction term has the main property of the bracket abstract in that it satisfies the property

(β) $(\lambda x . M)N$ reduces to $[N/x]M$,

where $[N/x]M$ is the result of substituting N for x in M, where the substitution is so defined that bound variables (such as the x in $\lambda x . M$) are changed to avoid collisions of bound variables. But there is one major difference with combinatory logic: the bracket abstract $[x]M$ is not a

primitive operation of the system, but is a metatheoretic abbreviation for a term in which there is no occurrence of x. On the other hand, x does occur in $\lambda x \, . \, M$, where it occurs bound. (There are no bound variables in combinatory logic.) Thus, in λ-calculus, the property

(ξ) If $M = N$, then $\lambda x \, . \, M = \lambda x \, . \, N$

is true by the definition of the system, whereas in combinatory logic

(ξ) If $X = Y$, then $[x]X = [x]Y$

only holds because of the combinatory axioms, and the corresponding property fails if $=$ is replaced by weak equality.

Another axiom of λ-calculus that is always assumed is

(α) If y does not occur free in M, $\lambda x \, . \, M = \lambda y \, . \, [y/x]M$.

A further axiom that is often added is

(η) If x does not occur free in M, then $\lambda x \, . \, Mx = M$.

It turns out that for λ-conversion, (ξ) plus (η) is equivalent to

(ζ) If x does not occur free in M and N, and if $Mx = Nx$, then $M = N$.

This is the extensionality property that follows in combinatory logic by the combinatory axioms. If the rule (η) is not postulated, the result is called the $\lambda\beta$-calculus; if it is the result is called the $\lambda\beta\eta$-calculus. (These names (the lower case Greek letters) are due to Curry.)

After Church extracted λ-calculus from his system of logic, the former system was used to prove some important results in logic: the existence of an unsolvable problem in elementary number theory [8], a constructive representation of ordinal numbers of the second number class [9], [10], and [41], the equivalence of definability of a numerical function in λ-calculus and being partial recursive [40], and what came to be called the Church-Rosser Theorem [11]. (The result on numerical functions was obtained by representing the natural number n by the term

$$\lambda x \, . \, \lambda y \, . \, \underbrace{x(x(\cdots(x\,y)\cdots))}_{n}.)$$

Curry, on the other hand, did not give up on the idea of using combinatory logic as a basis for a system of formal logic that would imply all of mathematics. Instead, he set out to study in detail the contradiction Kleene and Rosser had found in the system. He also found a slightly reduced set of combinatory axioms that made his combinatory equality equivalent to $\lambda\beta$-equality; his original ones were for an equality equivalent to $\lambda\beta\eta$-equality.

Curry spent most of the rest of the 1930s studying the contradiction of Kleene and Rosser, and the first result was the paper [19]. Later, he found a simpler contradiction, which is now known as *Curry's Paradox* [20]. On this basis, Curry set out a program to find consistent systems of illative combinatory logic. He proposed, in [21], three kinds of systems classified by the logical connectives that were taken as primitive:

1. Systems of *functionality*, called systems of type $\mathfrak{F}_1$, based on the primitive operator F of the theory of functionality. Here, F is defined by the elimination rule

   Rule F. $FXYZ, XU \vdash Y(ZU)$.

   Furthermore, FXY is supposed to represent what we now call the *type* $X \to Y$. Implication P and the universal quantifier Π can be defined in terms of F (and a category (or type) E which is a type of every term), as follows:

   $$PXY \equiv F(KX)(KY)I,$$
   $$\Pi X \equiv F(KE)(KX).$$

   (Curry used $\equiv$ to indicate identity as terms.) There were different types of functionality depending on which terms were allowed as types. In *basic* functionality, types were all formed from certain atomic types $\theta_1, \theta_2, \cdots$ by the operation forming $F\alpha\beta$ from α and β. On the other hand, there was *full, free* functionality in which any term could be a type. The rules included Rule F above and a rule

   Rule Eq. $X, X = Y \vdash Y$.

2. Systems of *restricted generality*, called systems of type $\mathfrak{F}_2$. In these systems, the primitive operator is what Curry called Ξ, where ΞXY represents $(\forall u)(Xu \supset Yu)$. The elimination rule for Ξ is

   Rule Ξ. $\Xi XY, XU \vdash YU$.

   This Ξ can be defined in terms of F: $\Xi XY \equiv FXYI$. Furthermore, P and Π can be defined in terms of Ξ and E:

   $$PXY \equiv \Xi(KX)(KY),$$
   $$\Pi X \equiv \Xi E.$$

   F can be defined in terms of Ξ by

   $$FXYZ \equiv [u](\Xi(Xu)(Y(Zu))).$$

3. Systems of *universal generality*, called systems of type $\mathfrak{F}_3$. In these systems, the primitive operators are P and Π. Ξ can be defined in terms of Π and P by

$$\Xi XY \equiv \Pi([u].P(Xu)(Yu)).$$

And then F can be defined as above.

Curry originally thought that these kinds of systems were of increasing strength. This ultimately turned out not to be the case.

But before Curry could do any further work on combinatory logic, the United States entered World War II, and at the end of the academic year 1941–42, Curry took a leave of absence from Penn State to do applied mathematics for the war effort.

However, just before he started his leave, he discovered the essential role of Schönfinkel's combinator S. He discovered that it was possible to define the bracket abstract by a simple induction on the structure of the terms as follows:

1. If a is any atomic term distinct from x, then $[x]a \equiv Ka$,

2. $[x]x \equiv I$,

3. If $[x]X$ and $[x]Y$ are defined, then $[x]XY \equiv S([x]X)([x]Y)$.

This result was only published after the war in [22].

After the war ended and Curry returned to academic life, it took him some time to get back to logic, since he was continuing some work in other areas he had done during the war. But in 1948, he attended the Tenth International Congress of Philosophy in Amsterdam, and while he was there he was approached by people associated with the new North-Holland Publishing Company to write a short (about 100 pages) monograph on combinatory logic. Curry realized immediately that he could not write such a short book on the subject, so instead he sent them [23], which was ready at the time. But he also started thinking about a book on combinatory logic. Since exposition would be important and Curry knew that he was not a very good expository writer, he felt the need for a collaborator. He had known of the work of Robert Feys as early as 1938, and in about 1947 to 1948, he saw a review by Kleene (in the *Journal of Symbolic Logic*) of [47]. Curry asked Robert Feys to be his co-author. He obtained a Fulbright grant to spend the year 1950–51 at Louvain to work with Feys, and let Feys take the lead on the exposition of the early chapters.

These early chapters introduce λ-calculus before combinatory logic, a trend in exposition that continues to this day. The book actually begins with two chapters explaining Curry's notion of formal system and

what he then called epitheory, which is really metatheory. Then comes the chapter introducing λ-calculus. There follows a long chapter proving the Church-Rosser Theorem for λ-calculus, the theorem which says that any two convertible terms have a term to which they both reduce. Then comes a chapter on the intuitive theory of combinators, which is followed by a chapter on the "synthetic theory of combinators"; the latter includes the analysis of "combinatorial completeness" (the definition of $[x]X$), the combinatory axioms, of which two versions are given, one equivalent to $\lambda\beta$-calculus and the other equivalent to $\lambda\beta\eta$-calculus, the equivalence to both forms λ-calculus, and finally a section on *strong reduction*, which was new work (see below). There follows a chapter called "Logistic Foundations", which introduces a theory of combinatory logic as rules, and then a chapter introducing *illative combinatory logic*. This introduction to illative combinatory logic includes a discussion of Curry's Paradox, and introduces the notion of functionality. Then come the last two chapters of the book, the first on basic functionality and the second on stronger theories of functionality.

The book also began a practice largely followed since then of printing Roman letters designating combinators in a sans-serif font, such as, F, K, S, B, C, W, and I Instead of F, K, S, B, C, W, and I. To make the translation closer to Curry's original, we have not followed this practice so far, but we will from now on in this introduction.

Curry began his new research for the book with the theory of functionality. In that theory, the statement that term X has type α is αX (thinking of the type as a predicate and the term as its subject). The axiom schemes are the following:

(FK) $F\alpha(F\beta\alpha)K$,

(FS) $F(F\alpha(F\beta\gamma))(F(F\,\alpha\beta)(F\alpha\gamma))S$.

The only two rules are the following:

    Rule F. $F\alpha\beta X, \alpha Y \vdash \beta(XY)$,

    Rule Eq. $X = Y, X \vdash Y$.

For type theorists, this can be read as K has type $\alpha \to (\beta \to \alpha)$ and S has type $(\alpha \to (\beta \to \gamma)) \to ((\alpha \to \beta) \to (\alpha \to \gamma))$.

The premises of a deduction were not restricted to assigning types to variables.

At first, Curry thought the full theory of functionality (in which any term can be a type) was consistent, and, in fact, he thought he knew how to prove it. In July, 1952, he started writing a section for the book on this consistency proof. But he was also writing up the basic theory of functionality (types are generated from atomic types by the forming of

F$\alpha\beta$ from α and β). Working on his idea for the consistency proof of the full theory led Curry to think more about reduction than he was used to. The only reduction he had was weak reduction generated by the rules for replacing KXY by Y and SXYZ by $XZ(YZ)$, and this reduction does not give the Church-Rosser property in relation to his equality, which we now call combinatory $\beta\eta$-equality; i.e., it is not true that if $X = Y$, then there is a term Z to which both weakly reduce.

As part of the section on basic functionality, Curry wrote of a correspondence between types of the form F$\alpha\beta$ and implication formulas of the form P$\alpha\beta$, or $\alpha \supset \beta$. He had been using this idea to give names to formulas involving only implication since the early 1930s, but he had only previously published it in [24, pp. 140–142]. In the book he was writing, [29, §9E] he showed how to use this correspondence to find proofs in a formulation of the implication fragment of intuitionistic logic whose only rule is modus ponens and whose axioms are the formulas corresponding to the types of K and S. This was the beginning of the idea of "propositions as types" or "formulas as types", and later several people, including W. A. Howard in [38], extended it to the other logical connectives and quantifiers.

In June, 1954, Curry began to realize that his idea for the consistency proof for full functionality seemed to imply that if a term X has a type in basic functionality then it has an irreducible form in some sense of reduction. But that sense in which it has an irreducible form is that the corresponding λ-term X_λ must have a normal form with respect to λ-reduction. Because he tended to think primarily in terms of combinatory logic rather than λ-calculus, he defined what he called *strong reduction*, what we now call *combinatory $\beta\eta$-reduction*. He later worked out the theory of this new reduction in some detail, and the result appeared in Chapter 6 as the new research of that chapter. For technical reasons, Curry needed to take I as a new primitive combinator, and so ever since the primitive combinators are usually included I, K, and S.

A bit later that July, he proved that the full theory of functionality is inconsistent. This proof appeared in [29, §10A3]. The proof used the Curry Paradox and the notion of propositions as types to prove that WWW has every possible term as a type (recall that W$xy = xyy$). In the terms we use today, this would be written as WWW : β, but Curry wrote it as β(WWW). He then let β be KX for an arbitrary term X, giving KX(WWW), which he then reduced to X, thus showing that in the full theory of functionality any term can be proved. That last inference would not be possible in current theories of typed λ-calculus or type assignment.

The writing of the book was finished in 1956, and the book appeared

in 1958. Curry was planning a second volume, which was to include work on restricted generality and universal generality, and he did publish in the late 1950s two papers on restricted generality, [25, 26]. But he also decided to write a graduate level textbook [27], mainly on Gentzian (i.e., Gentzen-style) proof theory, which he expected to want to use in the second volume.

In the early 1950s, Curry finally started to have graduate students after financial support became available at Penn State, all of whom worked on combinatory logic, and his first student, Edward J. Cogan, wrote a dissertation on set theory based on combinatory logic, [12]. Unfortunately, the system Cogan used turned out to be inconsistent, as shown in [58]. Curry took the blame for the contradiction since it resulted from a postulate he had suggested to Cogan. There followed three more graduate students, Kenneth Loewen (dissertation [44]), and Bruce Lercher (dissertation [43]), both of whom wrote on strong reduction, and Luis E. Sanchis (dissertation [49]), who wrote on functionality and type theory. Then came Jonathan P. Seldin (dissertation [51]), who wrote on illative combinatory logic in general, and then Martin W. Bunder (dissertation [4]), who wrote on set theory in combinatory logic. (More on these latter two below.) And finally, John A. Lever wrote a master's thesis with him in 1977.

In 1960, Curry became Evan Pugh Research Professor at the Pennsylvania State University, and was thus released from undergraduate teaching duties. This made it easier for him to do his research.

In 1964, Curry began work on the second volume of *Combinatory Logic*. He started working alone, since Robert Feys had died on April 13, 1961, but it soon became apparent to him that he needed other collaborators. J. Roger Hindley, who arrived at Penn State to assume a lectureship in September 1964 after finishing his dissertation [37], was invited to become a co-author in 1965, and Seldin was invited after finishing his dissertation [51] in 1968. The writing of the second volume was finished in 1970, and it appeared as [30] in 1972.

The book opens with a chapter on changes in pure combinatory logic since the appearance of [29] in 1958. Among the changes discussed is the problem of finding a combinatory reduction equivalent to $\lambda\beta$-reduction. Curry's original strong reduction is equivalent to $\lambda\beta\eta$-reduction, and part of the reason it is equivalent is that the rule (η) is built into the definition of the bracket abstract. Now, in $\lambda\beta$-calculus, there are some cases of (η) that are valid: If U is an abstract, say $\lambda y \,.\, V$, and if x does not occur free in V, then

$$Ux \equiv (\lambda y \,.\, V)x \rhd_\beta [x/y]V,$$

where $\rhd_\beta$ is $\lambda\beta$-reduction, and by a change of bound variables (rule

(α)), $[x/y]V$ converts to V. So to fully follow the definition of η-strong reduction, we would need to have $[x]Vx \equiv V$ as part of the definition of abstraction when V is β-equal to an abstraction term. But it is not decidable whether any term is β-equal to an abstraction term, and so such a definition of abstraction is not an algorithm. This problem of finding a combinatory reduction equivalent to $\lambda\beta$-reduction is not settled to this day; see [53].

The next chapter is on the foundations of illative theories. Included is a Gentzen-style L-system for $\mathcal{Q}$, a system in which the only operational constant is Q for equality. The next chapter is on combinatory arithmetic, which means the representation of natural numbers and numerical functions. The chapter includes a number of different ways of representing the natural numbers.

The next chapter is on functionality. This chapter consists of a number of extensions and revisions of the chapters at the end of [29]. Included are rules for a general separation of equality rules for the subjects (terms) and predicates (types), as is the case in all more modern typed combinatory and λ-calculi.

There follow two chapters, one on restricted generality and another on universal generality. It turned out that Curry's original idea that systems of universal generality are always stronger than those of restricted generality was wrong, and the systems are essentially equivalent. There are consistent systems of both kinds, but they are not really stronger than systems of first order logic; they are obtained by defining *canonical* terms which correspond to first-order formulas, and requiring that terms appearing in logical formulas be canonical. There was a desire to find *finite* formulations in which these restrictions to canonical terms could be introduced as rules of the system. So, for example, if there is a constant H which represents the predicate of being a proposition, the idea is that $\vdash HX$ should hold only if X is to be considered a proposition. In these terms, the ordinary deduction theorem for implication, also known as the rule for implication introduction, should take the form

$$\frac{\begin{array}{c}[X]\\Y\end{array} \quad HX \quad HY}{PXY}$$

Martin Bunder, in preparing his dissertation [4], proposed instead a rule of this form in which the only limitation is on the premise for the antecedent, so his version of this rule is

$$\frac{\begin{array}{c}[X]\\Y\end{array} \quad HX}{PXY}$$

Attempts were made to find consistent systems of this kind by finding axioms from which these deductive theorems could be proved. Unfortunately, all the proposed systems of this kind turned out to be inconsistent after the book was published; see [5, 6].

Curry's last proposal for a consistent system of this kind is in [28].

The book ends with a chapter on logical type theory, using the theory of functionality to construct the types.

At the beginning of the chapter on restricted generality, Curry introduced a notion of *generalized functionality*, a constant G which satisfies

Rule G. $\mathsf{G}XYZ, XU \vdash YU(ZU)$.

In generalized functionality, the type of the conclusion of the typing rule can depend on the argument as well as the type of the argument. Seldin wrote a paper on generalized functionality in the style of a chapter of this book, [52]. There are now many type theories based on this idea, but with a different notation; the type that Curry wrote as $\mathsf{G}AB$ is now usually written $(\Pi x : A \,.\, Bx)$. Typed systems based on this idea now include all the *pure type systems* [2, 57], including *Barendregt's cube* [1]. One of the pure type systems, and the strongest system in Barendregt's cube, is the *calculus of constructions*, [13], which is the basis for the proof assistant Coq [31]. Another proof assistant, HOL, is based on the type theory of Martin-Löf, [45]. For more on type theory in relation to combinatory logic and λ-calculus, see [39].

There have been other applications of Curry's combinatory logic, generalised functionality (and hence modern types) and propositions as types (PAT) in modern mathematics, computer science and linguistics. The literature is huge to list. In what follows we give a small number of examples. First, in linguistics, combinators à la Curry have been used to provide minimal grammars that are useful for describing universal natural language where for example it is shown that the combinator B can describe long distance sentence dependencies and the combinator W can be used for interpreting reflexive pronouns. Combinatory grammar and combinatory categorical grammar are thriving areas of research initiated by Mark Steedman and colleagues [55] whose starting point was that "the striking parallel between the particular set of combinators that is implicit in the grammar of English (and, by implication, other languages), and the systems of combinators that are used in certain highly efficient compilers for programming languages suggests that the reason natural language grammars take this form [..] maybe to do with the computational advantages of avoiding the use of bound variables." The Combinatory Categorial Grammar Site http://groups.inf.ed.ac.uk/ccg/ provides further details of this interesting application of Curry's combinators to natural language parsing and computation. Combinatory logic has also

played a crucial role in the design and implementation of programming languages. In 1977 David Turner started applying the normal graph reduction of Wadsworth [62] to Curry's SK combinators and reimplemented the St Andrew's Static language (SASL) he had developed earlier. Using combinatory logic, Turner showed that an applicative language, such as LISP, can be translated into a form where all bound variables have been removed and more importantly where the resulting code can be efficiently executed. Turner showed that for SK-combinator reduction, normal order reduction is optimal [60, 59, 61]. For details on the optimisation techniques and the pros and cons of the SK-combinators, see [54].

Part of the significance of Turner's result mentioned in the last paragraph is that in general, if one wants to implement λ-calculus using SK-combinators by using $[x]M$ to implement $\lambda x \,.\, M$, a problem arises because the standard definitions of $[x]$ are not efficient, and often blow up at inconvenient places. In 1989, Piperno [46] showed that Curry's original definition of $[x]$, from the dissertation and [16], is, with only a minor modification, more efficient that the other definitions known at the time.

As for the idea of PAT, we already mentioned the correspondence Curry made between types of the form $\mathsf{F}\alpha\beta$ and implication formulas of the form $\mathsf{P}\alpha\beta$, or $\alpha \supset \beta$ which he had been using to give names to formulas involving only implication. Curry showed how to use this PAT correspondence to find proofs in a formulation of the implication fragment of intuitionistic logic whose only rule is modus ponens and whose axioms are the formulas corresponding to the types of K and S. Howard [38] combined Curry's Propositions as Types correspondence with Tait's correspondence between cut elimination and β-reduction in order to treat proofs themselves as λ-terms [56], hence turning PAT into a proofs as terms as well as a propositions as types correspondence. This way, the typing judgement $\Gamma \vdash t : T$ which states that the term t has type T in context Γ corresponds to the logical judgement that in context Γ, proof t is a proof of Proposition T. The idea of PAT surfaces again in the work of N.G. de Bruijn in Automath [3], but it is clear that de Bruijn's work on PAT was independent of that of Curry and Howard and influenced by Heyting [35] (cf. [39]). PAT plays an important role in proof certification and proof carrying code, a code appear with its own proof of correctness so that correctness of specifications can be ensured. Not only Automath, but also other systems such as the Edinburgh Logical Framework (LF) [33] and Coq [31] use the PAT principle.

Curry's generalized functionality idea where the type of the conclusion of the typing rule can depend on the argument as well as the type of the argument has since been crucial in modern systems of logic and

computation where polymorphic and/or dependent types play a crucial role. We already mentioned Coq and LF which use both polymorphic and dependent types. Similarly, modern programming languages like Haskell, ML, and OCaml make heavy use of polymorphism. Incidentally, modern programming languages not only use Curry's generalised functionality, but also make sure terms are typed à la Curry (i.e., implicitly) rather than à la Church (i.e., explicitly). This means that instead of demanding that the user provides the types of variables explicitly as in the A of $\lambda x : A . B$, they let the compiler take that burden and allow the user to simply write $\lambda x . B$. The compiler will automatically deduce the type of the variable x. This is how languages like ML and Haskell work. Of course, as was said by Aristotle around 2500 years ago, given a problem and a solution to the problem, one can check that the solution indeed solves the problem. But, given a problem and asking for a solution may not produce an answer (the problem may not be solvable). In type theory, it was an open problem for over 23 years whether in the polymorphic λ-calculus implicitly typed à la Curry, type finding is decidable (i.e., whether the compiler can find the type $\lambda x . B$ which means finding the type A of x). Joe Wells showed in [63] that not only type finding, but also type checking are not decidable in the polymorphic λ-calculus. Of course ML is based on a decidable fragment of polymorphic types.

Curry's ideas continue to influence developments in mathematics, logic and computation. We hope that making his original thesis available in English will help make his ideas clearer.

BIBLIOGRAPHY

[1] H. P. Barendregt. Lambda calculi with types. In S. Abramsky, D. M. Gabbay, and T. S. E. Maibaum, editors, *Handbook of Logic in Computer Science*, volume 2, pages 117–309. Oxford University Press, 1992.

[2] Stefano Berardi. Towards a mathematical analysis of the Coquand-Huet calculus of constructions and the other systems of the Barendregt cube. Technical report, Department of Computer Science, Carnegie-Mellon University and Dipartimento di Matematica, Universita di Torino, 1988.

[3] N.G. de Bruijn. The mathematical language AUTOMATH, its usage and some of its extensions. In M. Laudet, D. Lacombe, and M. Schuetzenberger, editors, *Symposium on Automatic Demonstration*, pages 29–61, IRIA, Versailles, 1968. Springer Verlag, Berlin, 1970.

[4] M. W. Bunder. *Set Theory Based on Combinatory Logic*. PhD thesis, University of Amsterdam, 1969.

[5] M. W. Bunder. Some inconsistencies in illative combinatory logic. *Zeitschrift für mathematische Logik und Grundlagen der Mathematik*, 20:199–201, 1974.

[6] M. W. Bunder. The inconsistency of $\mathcal{F}_{21}^*$. *Journal of Symbolic Logic*, 41(2):467–468, June 1976.

[7] A. Church. A set of postulates for the foundation of logic. *Annals of Mathematics*, 33:346–366, 1932.

[8] A. Church. An unsolvable problem of elementary number theory. *American Journal of Mathematics*, 58:345–363, 1936.

[9] A. Church. The constructive second number class. *Bulletin of the American Mathematical Society*, 44:224–232, April 1938.

[10] A. Church and S. C. Kleene. Formal definitions in the theory of ordinal numbers. *Fundamenta Mathematicae*, 28:11–31, 1936.

[11] A. Church and J. B. Rosser. Some properties of conversion. *Transactions of the American Mathematical Society*, 39:472–482, 1936.

[12] E. J. Cogan. A formalization of the theory of sets from the point of view of combinatory logic. *Zeitschrift für Mathematische Logik und Grundlagen der Mathematik*, 1:198–240, 1955. PhD thesis, The Pennsylvania State University, 1955.
[13] Thierry Coquand and Gérard Huet. The calculus of constructions. *Information and Computation*, 76:95–120, 1988.
[14] H. B. Curry. An analysis of logical substitution. *American Journal of Mathematics*, 51:363–384, 1929.
[15] H. B. Curry. The universal quantifier in combinatory logic. *Annals of Mathematics*, 32:154–180, 1931.
[16] H. B. Curry. Apparent variables from the standpoint of combinatory logic. *Annals of Mathematics*, 34:381–404, 1933.
[17] H. B. Curry. Some properties of equality and implication in combinatory logic. *Annals of Mathematics*, 35:849–860, 1934.
[18] H. B. Curry. First properties of functionality in combinatory logic. *Tôhoku Mathematical Journal*, 41:371–401, 1936.
[19] H. B. Curry. The paradox of Kleene and Rosser. *Transactions of the American Mathematical Society*, 50:454–516, 1941.
[20] H. B. Curry. The inconsistency of certain formal logics. *Journal of Symbolic Logic*, 7:115–117, 1942.
[21] H. B. Curry. Some advances in the combinatory theory of quantification. *Proceedings of the National Academy of Sciences of the United States of America*, 28:564–569, 1942.
[22] H. B. Curry. A simplification of the theory of combinators. *Synthese*, 7:391–399, 1949.
[23] H. B. Curry. *Outlines of a Formalist Philosophy of Mathematics*. North-Holland Publishing Company, Amsterdam, 1951.
[24] H. B. Curry. *Leçons de Logique Algébrique*. Gauthier-Villars and Nauwelaerts, Paris and Louvain, 1952.
[25] H. B. Curry. The deduction theorem in the combinatory theory of restricted generality. *Logique et Analyse*, 3:15–39, 1960.
[26] H. B. Curry. Basic verifiability in the combinatory theory of restricted generality. In *Essays on the Foundations of Mathematics*, pages 165–189. Magnes Press, Hebrew University, Jerusalem, 1961.
[27] H. B. Curry. *Foundations of Mathematical Logic*. McGraw-Hill Book Company, Inc., New York, San Francisco, Toronto, and London, 1963. Reprinted by Dover, 1977 and 1984.
[28] H. B. Curry. The consistency of a system of combinatory restricted generality. *Journal of Symbolic Logic*, 38:83–92, 1973.
[29] H. B. Curry and R. Feys. *Combinatory Logic*, volume 1. North-Holland Publishing Company, Amsterdam, 1958. Reprinted 1968 and 1974.
[30] H. B. Curry, J. R. Hindley, and J. P. Seldin. *Combinatory Logic*, volume 2. North-Holland Publishing Company, Amsterdam and London, 1972.
[31] G. Dowek et al. The coq proof assistant version 5.6, users guide. Technical Report 134, INRIA, Le Chesney, 1991.
[32] G. Frege. *Begriffsschrift, eine der arithmetischen nachgebildete Formelsprache des reinen Denkens*. Nebert, Halle, 1879. Also in [34], pages 1–82.
[33] R. Harper, F. Honsell, and G. Plotkin. A framework for defining logics. In *Proceedings Second Symposium on Logic in Computer Science*, pages 194–204, Washington D.C., 1987. IEEE.
[34] J. van Heijenoort, editor. *From Frege to Gödel: A Source Book in Mathematical Logic, 1879–1931*. Harvard University Press, Cambridge, Massachusetts, 1967.
[35] A. Heyting. *Mathematische Grundlagenforschung. Intuitionismus. Beweistheorie*. Ergebnisse der Mathematik und ihrer Grenzgebiete. Springer Verlag, Berlin, 1934.
[36] D. Hilbert and P. Bernays. *Grundlagen der Mathematik*, volume I. Springer, Berlin, 1934.
[37] J. R. Hindley. *The Church-Rosser Property and a Result in Combinatory Logic*. PhD thesis, Newcastle upon Tyne, 1964.
[38] W. A. Howard. The formulae-as-types notion of construction. In J. Roger Hindley and Jonathan P. Seldin, editors, *To H. B. Curry: Essays on Combinatory Logic, Lambda Calculus and Formalism*, pages 479–490. Academic Press, New York, 1980. A version of this paper was privately circulated in 1969.
[39] F. Kamareddine, T. Laan, and R. Nederpelt. *A Modern Perspective on Type Theory*. Kluwer, Dordrecht, Boston, London, 2004.
[40] S. C. Kleene. λ-definability and recursiveness. *Duke Mathematical Journal*, 2:340–353, 1936.
[41] S. C. Kleene. On notation for ordinal numbers. *Journal of Symbolic Logic*, 3(4):150–155, December 1938.
[42] S. C. Kleene and J. B. Rosser. The inconsistency of certain formal logics. *Annals of Mathematics*, 36:630–636, 1935.
[43] B. Lercher. *Strong reduction and recursion in combinatory logic*. PhD thesis, The Pennsylvania State University, 1963.

[44] K. Loewen. *A study of strong reduction in combinatory logic*. PhD thesis, the Pennsylvania State University, 1962.
[45] P. Martin-Löf. An intuitionistic theory of types: Predicative part. In H. E. Rose and J. C. Shepherdson, editors, *Logic Colloquium '73*, pages 73–118. North-Holland Publishing Company, Amsterdam, 1975.
[46] Adolfo Piperno. Abstraction problems in combinatory logic: A compositive approach. *Theoretical Computer Science*, 66:27–43, 1989.
[47] R.Feys. La technique de la logique combinatoire. *Revue philosophique de Lourain*, 44:74–103, 237–270, 1946.
[48] J. B. Rosser. A mathematical logic without variables, Part II. *Duke Mathematical Journal*, 1:328–355, 1935.
[49] L. E. Sanchis. *Normal combinations and the theory of types*. PhD thesis, The Pennsylvania State University, 1963.
[50] M. Schönfinkel. Über die Bausteine der mathematischen Logik. *Mathematische Annalen*, 92:305–316, 1924. English translation, "On the building blocks of mathematical logic," in Jean van Heijenoort, editor, *From Frege to Gödel: A Source Book in Mathematical Logic, 1879–1931*, pages 355–366, Harvard University Press, Cambridge, MA and London, 1967.
[51] J. P. Seldin. *Studies in Illative Combinatory Logic*. PhD thesis, University of Amsterdam, 1968.
[52] J. P. Seldin. Progress report on generalized functionality. *Annals of Mathematical Logic*, 17:29–59, 1979.
[53] J. P. Seldin. The search for a reduction in combinatory logic equivalent to $\lambda\beta$-reduction. *Theoretical Computer Science*, 412:4905–4918, 2011.
[54] Simon Peyton-Jones. *The Implementation of Functional Programming Language*. Prentice-Hall International Series in Computer Science, 1987.
[55] M. Steedman. Combinators and grammars. In Deirdre Wheeler Richard T. Oehrle, Emmon Bach, editor, *Categorial Grammars and Natural Language Structures*, volume 32 of *Studies in Linguistics and Philosophy*, pages 417–442. Springer, 1988.
[56] W.W. Tait. Infinitely long terms of transfinite types. In Crossley and Dummett, editors, *Formal Systems and recursive Functions*. North-Holland, 1965.
[57] J. Terlouw. Een nadere bewijstheoretische analyse van GTS's. Technical report, Department of Computer Science, Catholic University, Toernooiveld 1, 6525 ED Nijmegen, The Netherlands, 1989.
[58] R. Titgemeyer. Über einen Widerspruch in Cogans Darstellung der Mengenlehre. *Zeitschrift für Mathematische Logik und Grundlagen der Mathematik*, 7:161–163, 1961.
[59] D. Turner. Another algorithm for bracket abstraction. *Symbolic Logic*, 44(2):267–270, 1979.
[60] D. Turner. A new implementation technique for applicative languages. *Software-Practice and Experience*, 9(1):31–49, 1979.
[61] D. Turner. Some history of functional programming languages. *invited talk at TFP 2012, St Andrews university*, 2012.
[62] C.P. Wadsworth. *The Semantics and Pragmatics of lambda calculus*. PhD thesis, University of Oxford, 1971.
[63] J. B. Wells. Typability and type-checking in the second-order lambda-calculus are equivalent and undecidable. In *Proceedings of the Ninth Annual Symposium on Logic in Computer Science (LICS '94), Paris, France, July 4-7, 1994*, pages 176–185. IEEE Computer Society, 1994.
[64] A. N. Whitehead and B. Russell. *Principia Mathematica*. Cambridge University Press, Cambridge, England, 1910–1913. Second edition, 1925–1927.

Foundations of Combinatory Logic, by H. B. Curry:
Translation into English
FAIROUZ KAMAREDDINE AND JONATHAN P. SELDIN

ABSTRACT. This is a translation of H. B. Curry's "Grundlagen der kombinatorischen Logik", *Amer. J. Math.* 52 (1930) 509–536, 789–834. The first part of the translation, sections A–C of Chapter 1, was done during a visit by Seldin to Oxford, England in January and February, 1979, with the help of Gerhard Jäger and Karl Gosejacob, who were graduate students there at the time. The rest of the translation is new. Throughout, we tried to be as faithful as possible to Curry's original presentation. At the same time, we have made use of all information we could find by Curry, mostly from marginal notes in his copy of the dissertation, indicating corrections to the original German.

The footnotes are Curry's. The endnotes are notes by the translators, and literature references in these endnotes refer to the Bibliography that follows "The Work of H. B. Curry on Combinatory Logic".

The practice of writing combinators in sans-serif characters, such as I, K, and S, began with the publication of [29] in 1958. We are not following that practice in this translation.

In the original, citations to the literature were not indicated the way we indicate them today, but were given only in the footnotes. The references listed here are only those of the translators.

We would like to thank Martin Bunder and Roger Hindley for their helpful comments and suggestions.

Grundlagen der kombinatonschen Logik.

TEIL I.

von H. B. Curry.

INHALTSÜBERSICHT.

KAPITEL I. Allgemeine Grundlagen.
    Abschnitt A. *Einleitung.*
    Abschnitt B. *Einige philosophische Betrachtungen.*
    Abschnitt C. Das *Grundgerüst.*
    Abschnitt D. *Die Eigenschaften der Gleichheit.*

KAPITEL II. Die Lehre der Kombinatoren.
    Abschnitt A. *Einleitung.*
    Abschnitt B. *Grundlegende Definitionen und Sätze.*
   *Abschnitt C. *Darstellung von Kombinationen durch Kombinatoren.*
    Abschnitt D. *Reguläre Kombinatoren.*
    Abschnitt E. *Eigentliche Kombinatoren.*

KAPITEL I. ALLGEMEINE GRUNDLAGEN.

A. Einleitung.

Im Anfang jeder mathematisch-logischen Untersuchung setzt man eine gewisse Menge von Kategorien voraus, die als ein Teil des irreduziblen Minimums von Kenntnis, womit man anzufangen hat, betrachtet werden sollen, Als solche Kategorien gelten gewöhnlich Aussagen und Aussagefunktionen von verschiedenen Ordnungen und Stufen. Diese Kategorien müssen weiterhin nicht nur als rein formale Begriffe vorausgesetzt werden, sondern es muss zuweilen einen Sinn haben, dass ein gegebener Gegenstand zu dieser oder jener Kategorie gehört. Mit anderen Worten, sie müssen eine inhaltliche Eigenschaft besitzen, nämlich die, dass sie Kategorien sind; und so sind sie das, was ich schon nicht-formale Grundbegriffie genannt habe.†

    ∗ Die Abschnitte C, D, E werden in Teil II erscheinen.

    † In einer Abhandlung, " An Analysis of Logical Substitution," *American Journal of Mathematics*, Bd. 51 (Juli, 1929), S. 363-384.

Foundations of Combinatory Logic.

PART I

by H. B. Curry.

Table of Contents

Chapter I.   General Foundations.
              Section A. Introduction.
              Section B. Some Philosophical Remarks.
              Section C. The Primitive Frame.
              Section D. The Properties of Equality.
Chapter II. The Theory of Combinators.
              Section A. Introduction.
              Section B. The Basic Definitions and Theorems.
              Section C. The Representation of Combinations
                      via Combinators.
              Section D. Regular Combinators.
              Section E. Pure Combinators.

Chapter I. General Foundations

A. Introduction

At the beginning of any mathematical-logical inquiry one presupposes a certain set of categories (*Kategorien*), which is considered as a part of the irreducible minimum of knowledge with which one begins. As such categories, one usually has propositions (*Aussagen*) and propositional functions (*Aussagefunktionen*) of various orders and degrees. These categories, moreover, are not to be merely presupposed as purely formal ideas, but it must sometimes make sense that a given object belongs to this or that category. In other words, they must possess a contensive[1] (*inhaltlich*) property, namely that they are categories; and so they are what I have called non-formal primitive ideas.[†]

[†]In a paper, "An Analysis of Logical Substitution," *American Journal of Mathematics*, vol. 51 (July, 1929), pp. 363–384.

Die mit diesen Kategorien verbundene vorausgesetzte Kenntnis reicht aber weit über diese Eigenschaft hinaus. In der Tat müssen wir neben den Kategorien auch gewisse Verknüpfungen annehmen—die natürlich wieder nicht-formale Grundbegriffe sind, indem sie Verfahren angeben, wonach zwei oder mehrere Gegenstände kombiniert werden können, um einen neuen Gegenstand hervorzubringen. Unter diesen Verknüpfungen gehen uns hier alle Arten von Substitutionsprozessen an. Davon muss man die folgenden Eigenschaften anfänglich erkennen: erstens diejenigen, welche uns die Möglichkeit eines Substitutionsprozesses kundgeben, wenn nur die betreffenden Gegenstände zu gewissen Kategorien gehören; zweitens die, woraus wir unter Umständen schliessen, dass zwei verschiedene Substitutionsprozesse dasselbe Resultat liefern*; drittens diejenigen, welche die Kategorie des Ergebnisses bestimmen. Diese Kategorien, Verknüpfungen und Eigenschaften bilden daher eine vorausgesetzte Lehre von beachtenswerter Verwicklung; und diese Lehre möchte ich die *Urlogik* der betreffenden Theorie mennen.

Trotz des fundamentalen Charakters dieser Urlogik für irgendeine existierende mathematische Theorie der Logik, haben doch einige der wichtigen Probleme schon in ihr ihren wesentlichen Ursprung. Ich betrachte hier die zwei folgenden:

1) Die Vereinfachung der Grundlagen. Das Wesen dieses Problems scheint mir darin zu liegen, dass man die gesamte Mathematik und Logik auf ein Minimum von Urkenntnis aufbauen will; oder genauer, dass man sie in ihre Elemente zerlegen will, damit diese Urkenntnis mit schärfster Klarheit und Deutlichkeit ausgeprägt wird. Aus diesem Grunde strebt jedermann vermutlich danach, die Anzahl der Grundbegriffe und Axiome zu vermindern; und zwar sind einige Logiker so weit in dieser Richtung gegangen, dass sie nicht davor zurückschrecken, Allgemeinheit, Vollständigkeit und sogar Exaktheit diesem Streben zu opfern.† Allein für die Logik ist die Anzahl der Grundbegriffe und Axiome dieser Art eine Betrachtung von geringerer Wichtigkeit. Denn was macht es eigentlich, dass wir zwei oder drei Grundbegriffe ausschalten, wenn schon in der Urlogik selbst unendlich viele Grund-

---

*Vgl. Kapitel II unten; auch für einfache Beispiele den Teil II meiner oben zitierten Abhandlung.

† Z. B. müssen wir in den *Prinaipia Mathematica*, Implikation und den bestimmten Artikel als Grundbegriffe entbehren. Die dort gegebenen Definitionen dieser Grundbegriffe führen zu den paradoxen Resultaten. In der Tat ist dort der König von Frankreich von dem König von Frankreich verschieden; und zwischen den zwei Aussagen: 1) "hatten alle Menschen drei Hände, so würde auch Bismarck drei Hände haben," 2) "hätten alle Menschen drei Hände, so wäre der Mond aus grünem Käse aufgebaut," wird kein Unterschied gemacht.

The presupposed knowledge associated with these categories goes far beyond this property. In fact we must assume in addition to the categories certain modes of combination (*Verknüpfungen*) – which are naturally more non-formal primitive ideas, since they give procedures by which two or more objects can be combined to form a new object. Among these modes of combination we include all kinds of substitution processes. Among these one must recognize from the beginning the following properties: first those which give us the possibility of a substitution process when at least the objects in question belong to certain categories; second, those from which we conclude in certain circumstances that two different substitution processes yield the same result;* third those which determine the category of the results. These categories, modes of combination, and properties form therefore a presupposed theory of significant complication; and I shall call this the *prelogic* (*Urlogik*) of the theory in question.

In spite of the fundamental character of this prelogic for any existing mathematical theory of logic, some important problems have their essential origin in it. I consider here the following two:

1) The simplification of the foundations. The nature of this problem appears to me to be that one wants to construct all mathematics and logic on a minimum of primitive knowledge; or more exactly one wants to analyze them into their elements in order to express this primitive knowledge with the sharpest clarity and distinctness. For this reason I suppose everybody tries to reduce the number of primitive concepts and axioms; and some logicians have gone so far in this direction that they do not shrink from sacrificing generality, completeness, and accuracy.† However, the number of primitive concepts and axioms of this kind is a consideration of little importance for logic. For what difference does it really make that we remove two or three primitive concepts when infinitely many primitive concepts and rules, i.e. not formal

---

*Cf. Chapter II below; also for simple examples Part II of my paper cited above.

†E. g. we must do without implication and the definite article as primitive concepts in the *Principia Mathematica*. The definitions of these primitive concepts there lead to paradoxical results. In fact, there the King of France is different from the King of France; and no distinction is made between the two propositions 1) "If all men had three hands, Bismarck would have three hands," and 2) "If all men had three hands, the moon would be made out of green cheese."

begrifle und Regeln, und zwar nicht formale, vorhanden sind? Die Vereinfachung der Grundlagen soll also mit der Vereinfachung der Urlogik beginnen.

2) Die Beseitigung der Antinomien. Um uns die Verwandtschaft von diesen mit der Urlogik zu vergegenwärtigen, betrachten wir z. B. das Russellsche Paradoxon, das in der folgenden Weise aufgefasst werden kann: F sei diejenige Eigenschaft von Eigneschaften, für welche
$$F(\phi) = \text{nicht } \phi(\phi);$$
dann ist $\quad F(F) = \text{nicht } F(F).$

Behauptet man, dass $F(F)$ eine Aussage sei, wo eine Aussage als etwas, das entweder wahr oder falsch ist, definiert wird, so hat man sofort einen Widerspruch. Aber wir können den Widerspruch allerdings dadurch vermeiden, dass wir in Abrede stellen, dass $F(F)$ zu der Kategorie der Aussagen, oder F selbst zu der der Eigenschaften gehöre. Gerade hier stossen wir auf einen Lehrsatz der Urlogik. Und zwar sind es diese Antinomien, die uns zwingen, eine entwickeltere Urlogik uns anzueignen, als sonst begreiflich wäre. Es mag also sein, dass ein tieferes Studium der Urlogik Licht in dieses dunkle Gebiet der Antinomien verbreiten wird.

Diese Umstände haben mir die Anregung gegeben, jene Urlogik mathematisch zu behandeln. D. h. geuaner: eine abstrakte oder formale Theorie aufzubauen, welche auf eine sehr einfache Urlogik gegründet wird, und durch welche die Fragen, die man in der gewöhnlichen anschaulichen Urlogik zu stellen pflegt, durch symbolische Beweisführung beantwortet werden können. In dieser Abhandlung beabsichtige ich eine solche Theorie zu begründen. Weil die betreffendem Fragen einen wesentlich kombinatorischen Charakter haben, habe ich die Theorie *kombinatorische Logik* genannt.

Diese kombinatorische Logik wird wohl fähig sein als Grundlage einer abstrakten Theorie der gesamten Logik und Mathematik, einschliesslich Funktionen (Prädikaten, Relationen) von beliebig vielen Variablen, zu dienen. In der Tat bin ich der Ueberzeugung, dass eine solche Theorie durch Hinzufügen von endlich vielem formalen Grundbegriffen und Axiomen zu dem unten gegebenen Grundgerüst * gegründet werden kann. Weil in einer so eingerichteten Theorie das Grundgerüst überhaupt nur endlich viele Bestandteile hat, und weil ferner die Regeln nur ungefähr denselben Grad von Zu-

---

* Dieses Wort ist eine Uebersetzung der englischen Phrase "primitive frame," die ich in der oben erwähnten Abhandlung definiert habe. Es bedeutet die Gesamtheit der vorausgesetzten Grundbegriffe, Axiome und Regeln.

are already present in the prelogic? The simplification of foundations should therefore begin with the simplification of the prelogic.

2) The elimination of the paradoxes. In order to see the relationship of these paradoxes to the prelogic, we consider e.g. the Russell Paradox, which can be stated in the following way: let F be that property of properties ϕ for which

$$F(\phi) = \text{not } \phi(\phi);$$

then

$$F(F) = \text{not } F(F).$$

If one maintains that $F(F)$ is a proposition, where a proposition is defined as something that is either true or false, then one has a contradiction. But we can avoid the contradiction by denying that $F(F)$ belongs to the category of propositions or F itself to that of properties. Just here we run into a theorem of the prelogic. And it is these paradoxes that force us to adopt a more complex prelogic than would otherwise be conceivable. It may be, therefore, that a deeper study of the prelogic will spread light in the dark territory of the paradoxes.

These particulars have suggested to me treating this prelogic mathematically. I.e. more exactly: constructing an abstract or formal theory, which will be based on a very simple prelogic and through which the questions that one is accustomed to place in the usual intuitive prelogic can be answered by symbolic reasoning. In this work I intend to establish such a theory. Because the relevant questions have essentially a combinatory character, I call the theory *combinatory logic*.

This combinatory logic will be capable of serving as a foundation for an abstract theory of all logic and mathematics, including functions (predicates, relations) of arbitrarily many variables. In fact I am convinced that such a theory can be constructed by the addition of finitely many formal primitive concepts and axioms to the primitive frame (*Grundgerüst**) given below. Because in a theory so arranged the primitive frame has overall only finitely many constituents, and because further the rules have only about the same degree of complexity as the well known rule

---

*This word is a translation of the English phrase "primitive frame", which I have defined in the paper cited above. It means the collection of the postulated primitive concepts, axioms and rules.

sammengesetztheit wie die wohlbekannte Schlussregel haben, so wird dabei ein wesentlicher Fortschritt gegen das erste oben erwähnte Problem erzielt. Was das zweite betrifft, kann ich zunächst darüber nichts aussagen.

Das in dieser Abhandlung ausgeführte Problem ist doch nur ein Teil der kombinatorischen Logik überhaupt. Dieser Teil ist die Analyse der Substitutionsprozesse, formell betrachtet und abgesehen von den Kategorien, wozu die Gegenstände gehören; d. h. ich will die zweite der drei im zweiten Paragraphen erwähnten Arten von Eigenschaften untersuchen.*

Der Weg zu dieser Analysis ist von M. Schönfinkel † angedeutet worden. Er versteht zunächst eine Funktion in einem etwas neuen Sinne, nämlich als eine Zuordnung eines Funktionswerts zu jedem Element eines Argumentbereiches, wo sowohl der Funkionswert, als auch das Argument eine Funktion sein kann. Er will dann eine Funktion von n Variablen (im gewöhnlichen Sinne) als eine Funktion (im neuen Sinne) betrachten, deren Funktionswert für das Argument x eine Funktion (im gewöhnlichen Sinne) von $(n-1)$ Variablen ist, und zwar diejenige, welche durch Einsetzung von x in die erste ‡ Leerstelle der ursprünglichen Funktion entsteht. Mit dieser Auffassung verknüpft er die folgende Bezeichnungsweise. Leerstellen gebraucht er nicht, sondern er bezeichnet Funktionen wie andere Dinge mit einfachen Buchstaben (oder Gruppen von Buchstaben). Den Funktionswert einer Funktion für ein gegebenes Argument bezeichnet er weiterhin damit, dass er das Zeichen für das Argument gleich rechts von dem für die Funktion geltenden schreibt. Zur Erläuterung dieser Ideen schreibe ich gleich unten links, einige Ableitungen von einer Funktion f, die wir im gewöhnlichen Sinne als eine Funktion von drei Variablen begreifen, und rechts dieselben Gegenstände in etwas anschaulicherer Schreibweise:

$$\begin{array}{ll} f & f(-_1, -_2, -_3) \\ fx & f(x, -_1, -_2) \\ (fx)y & f(x, y, -_1) \\ ((fx)y)z & f(x, y, z) \end{array}$$

Schönfinkel setzt weiter fest, dass in einem Ausdruck wie $\cdots (((fx_1)x_2)x_3) \cdots$ die die letzten Seiten mit umfassenden Klammern wegbleiben dürfen.

---

* Die erste Art wird auch erledigt. In der Tat sind diese Eigenschaften nach den Betrachtungen von B1 (unten) ganz trivial.

† *Mathematische Annalen*, Bd. 92 (1924), S. 305-316.

‡ Es ist allerdings hier vorausgesetzt, dass die Leerstellen in einer durch die Funktion selbst bestimmten Weise numeriert sind. Dieselbe Annahme liegt doch der Schönfinkelschen Darstellung zugrunde. Vgl. unten II, A.

of inference,[2] clear progress is attained concerning the first problem stated above. Concerning the second I can say nothing further at present.

The problem taken up in this work is only a part of combinatory logic in general. This part is the analysis of substitution processes, considered formally and apart from the categories to which the objects belong; i.e. I will investigate the second of the three kinds of properties given in the second paragraph.*

The way to this analysis has been indicated by M. Schönfinkel.[†] First he understands a function in a somewhat new sense, namely as an association of a functional value (*Funktionswert*) to each element of the domain of the arguments (*Argumentbereich*), where both the functional value as well as the argument may be a function. He then considers a function of n variables (in the usual sense) as a function (in the new sense) in which the functional value for the argument x is a function (in the usual sense) of $(n-1)$ variables, and in particular the one obtained by the substitution of x in the first argument place[3‡] (*Leerstelle*) in the original function.[4] With this interpretation he combines the following notation. He does not use blanks, but he denotes functions as well as other things with simple letters (or groups of letters). The functional value of a function for a given argument he denotes by writing the sign for the argument just to the right of that for the function. To explain this idea I write below on the left some expressions (*Ableitungen*) based on a function f, that we understand in the usual sense as a function of three variables, and on the right the same objects in a somewhat clearer notation:

$$\begin{array}{ll} f & f(-_1, -_2, -_3) \\ fx & f(x, -_1, -_2) \\ (fx)y & f(x, y, -_1) \\ ((fx)y)z & f(x, y, z) \end{array}$$

Schönfinkel further adopts the convention that in an expression like $\cdots(((fx_1)x_2)x_3)\cdots$ the parentheses may be omitted according to the principle of association to the left.

---

*The first will also be settled. In fact these properties are by the considerations of B 1 (below) entirely trivial.

[†] *Mathematische Annalen*, Vol. 92 (1924), pp. 305–316.

[‡] It is presupposed here that the blanks are enumerated in a way determined by the function itself. The same assumption lies at the basis of Schönfinkel's description. Cf. II A below.

Um nun etwas allgemeinere Ausdrücke darzustellen, setzt er einige bestimmte Funktionen fest. Diese möchte ich hier mit den Buchstaben I, K, B, C, S bezeichnen.* Sie sind dadurch definiert, dass für beliebige Gegenstände x, y, z die folgenden Definitionsregeln gelten:

(1)
$$\begin{aligned} Ix &= x \\ Kxy &= x \\ Bxyz &= x(yz) \\ Cxyz &= xzy \\ Sxyz &= xz(yz). \end{aligned}$$

Vermöge dieser Funktionen ist er in der Lage, andere Ableitungen einer Funktion von drei Variabeln ohne Leerstelle darzustellen, etwa

$$\begin{array}{ll} Cfx & f(-_1, x, -_2) \\ C(BCf)xy & f(y, -_1, x) \\ Bfg & f(g(-_1)) \end{array}$$

u. s. w. In der Tat gilt nun der allgemeine Satz: Wenn irgendein logischer Ausdruck vorhanden ist, der aus gewissen Konstanten $u_1, u_2, \cdots u_m$ und numerierten freien Variablen zusammengesetzt ist, so kann er in der Schönfinkelschen Schreibweise in der Form

(2) $\qquad Yu_1, u_2 \cdots u_m$

wo Y eine Zusammensetzung von nur I, K, B, C, S ist, dargestellt werden. Diese Darstellung bedeutet, dass, wenn man nach (2) die Variablen in ihrer natürlichen Reihenfolge hinzufügt, und dann die Funktionen I, K u. s. w. durch die Definitionsregeln (1) eliminiert, der resultierende Ausdrück immer mit dem zuerst gegebenen identisch ist.–Dieser Satz ist von Schönfinkel nicht bewiesen, aber er enthält vom hier vorliegenden Gesichtspunkte aus das Wesen seiner Gedanken. Der Satz ist hier allerdings etwas roh dargelegt; genauere Sätze werden aber unten im formalen Teile streng bewiesen.

Nun macht Schönfinkel Anspruch darauf, dass er die Begriffe Variable, Aussage und Aussagefunktion aus der Logik entfernt habe. In dieser Hinsicht sind aber folgende Betrachtungen zu bemerken. Erstens: das Y des eben dargelegten Satzes ist doch nicht eindeutig bestimmt; z. B. sind *BCCf* und *If* in dem Sinne identisch, dass sie denselben Ausdruck darstellen. Aber diese Identität ist keinesweges aus den Definitionsregeln zu beweisen; zu diesem Behuf muss man vielmehr die Variablen tatsächlich hinzufügen. Infolge-

---

* Schönfinkel hat diese bzw. mit I, C, Z, T, S bezeichnet.

Now in order to describe somewhat more general expressions, he fixes some particular functions. I will denote these here with the letters I, K, B, C, S.* They are so conceived that for any objects x, y, z the following definitional rules hold:

(1)
$$\begin{aligned} Ix &= x \\ Kxy &= x \\ Bxyz &= x(yz) \\ Cxyz &= xzy \\ Sxyz &= xz(yz). \end{aligned}$$

By virtue of these functions he is in the position to represent other constructions of a function of three variables without blanks, such as

$$\begin{array}{ll} Cfx & f(-_1, x, -_2) \\ C(BCf)xy & f(y, -_1, x) \\ Bfg & f(g(-_1)) \end{array}$$

etc. In fact, the following general theorem holds: if any logical expresion exists that is built up from certain constants $u_1, u_2, \cdots, u_m$ and numbered free variables, then it can be represented in the notation of Schönfinkel in the form

(2) $\quad Yu_1u_2 \cdots u_m$

where Y is a combination of only I, K, B, C, S.[5] This representation means that if one places after (2) the variables in their natural order and then eliminates the functions I, K, etc. by the definitional rules (1), the resulting expression is always identical with the first given.–This theorem is not proved by Schönfinkel, but it is entailed by the viewpoint given above from the essence of his thought. To be sure the theorem is explained here somewhat crudely; but more exact theorems will be rigorously proved below in the formal part.

Now Schönfinkel claims that he has removed from logic the concepts of variable, proposition, and propositional function. But in this respect the following remarks must be made. First: the Y of the theorem explained above is not uniquely determined; E.g., $BCCf$ and If are identical in the sense that they represent the same expression. But this identity can in no way be proved from the definitional rules; for this purpose one must in fact add variables. Consequently,

---

*Schönfinkel denotes these respectively by I, C, Z, T, S.

dessen sind doch die Variablen, obgleich sie in den Formeln selbst nicht erscheinen, mit allen ihren assoziierten Grundbegriffen noch unvermeidbar. Zweitens: Die Begriffe Aussage und Aussagefunktion sind auch nur scheinbar eliminiert. Denn jede Funktion hat ihren Argumentbereich, und wie definiert man diesen, wenn man jene Begriffe nicht versteht? Die algemeinen Sätze der Logik gelten ja gewöhnlich nur, wenn die Variablen zu bestimmten Kategorien gehören. Z. B. ist der Satz von der Identität $A \to A$, nur wahr, wenn A eine Aussage ist. Weil Schönfinkel keineswegs gezeigt hat, wie die Einführung jener Grundbegriffe zu vermeiden ist, und weil er sie nicht aus anderen definieren kann, so hat er seinen Anspruch nicht gerechtfertigt. Er hat in der Tat nur eine neue und unbequemere Schreibweise gewonnen.

Jedoch führen diese Gedanken Schönfinkels zur Lösung des oben vorgelegen Problems. Denn die Funktionen B, C, u. s. w. bilden die Elemente, in die die Substitutionsprozesse sich zerlegen lassen. (Ich werde nachher genau zeigen, wie der allgemeinste solche Prozess sich aus diesen Funktionen zusammensetzen lässt). Das Wesen des Problems liegt daher in den im ersten Teile des letzten Absatzes besprochenen Identitäten. Es lässt sich wohl auf die folgende Aufgabe zurückführen: das Grundgerüst so festzustellen, dass jede dieser Identitäten rein formal bewiesen werden kann. Diese Aufgabe wird unten vollständig gelöst,[*] und zwar so, dass 1) die Funktionen $B, C \cdots$ nur als rein formale Begriffe auftreten, und 2) die Variablen in den formalen Beweisen nicht zu erscheinen brauchen. Der Beweis, dass jede solche Identität aus dem Grundgerüst ableitbar ist, macht das Hauptergebnis des zweiten Kapitels aus. Dort wird auch auseinandergesetzt, wie dies die Lösung des früheren Problems liefert.

Diese Untersuchung habe ich immer in Rücksicht auf eine allgemeine Theorie der Logik durchgeführt. Dieses Ergebnis betrachte ich nur als eine Vorbereitung zu einer weiteren Theorie, die ich später fortzuführen hoffe.

B. Einige Philosophische Betrachtungen.

Bevor ich zum Aufbau der formalen Theorie übergehe, mochte ich hier einige philosophische Vorbemerkungen darlegen, die für die Theorie massgebend gewesen sind. Diese betreffen die logische Auffassung, welche der Theorie vorangeht und durch die Ausdeutung der letzteren heranreift. Mit der formalen Theorie als solcher haben sie natürlich nichts zu tun; und der Leser, der sich nicht dafür interessiert, kann sie ruhig überspringen.

---

[*] Einige Anfänge dazu befinden sich in meiner früheren Abhandlung. Die vorliegende Behandlung ist allgemeiner und in sich vollständig; ich weise auf die frühere nur für gewisse Einzelheiten hin.

variables, with all their associated primitive concepts, are still unavoidable, although they do not appear in the formulas themselves. Second: The concepts of proposition and propositional function are also only apparently eliminated. For every function has its domain of arguments, and how does one define this if one does not understand that concept? The general theorems of logic usually hold only when the variables belong to definite categories. E.g., the theorem of identity $A \to A$ is only true when A is a proposition. Because Schönfinkel has in no way shown how the introduction of the other fundamental concepts is to be avoided, and because he cannot define them from others, he has not justified his claim. In fact he has achieved only a new and inconvenient notation.

Yet these thoughts of Schönfinkel do lead to a solution of the problem stated above. For the functions B, C, etc. make up the elements out of which the substitution process may be analyzed. (I shall show later on exactly how the most general such process may be built up from these functions.) The essence of the problem lies in the identities mentioned in the first part of the last paragraph. It can easily be reduced to the following problem: to formulate the primitive frame so that each of these identities can be proved purely formally. This problem will be completely solved* below in such a way that 1) the functions $B, C, \cdots$ appear only as purely formal concepts, and 2) variables need not appear in formal proofs. The proof that every such identity is derivable from the primitive frame is the main result of the second chapter. It will also be shown there how this leads to a solution of the earlier problem.

I have carried out this work always in consideration of a general theory of logic. I consider this result only as a preparation for a broader theory, which I hope to carry on with later.

B. Some Philosophical Remarks

Before I go over the construction of the formal theory, I will set down here some philosophical remarks which are important for the theory. These concern the logical conception which precedes the theory and becomes manifest through the interpretation of the latter. With the formal theory as such they have, of course, nothing to do; and the reader who is not interested in them is welcome to skip them.

---

*Some beginnings for this can be found in my earlier paper. The present treatment is more general and self contained; I refer to the earlier one only for certain details.

§1. *Sinnlose Begriffe.* In der gewöhnlichen Logik kommen gewisse Gegenstände (Begriffe, Dinge) vor, die man als sinnlos zu titulieren pflegt. Ich möchte nun gerade fragen, was dies bedeutet. Man darf wohl behaupten, dass ein Wort oder ein Zeichen in bezug auf eine Sprache noch nicht definiert ist. Z. B. ist das Wort "cow" auf deutsch sinnlos, doch gelegentlich anf englisch sinnvoll. Aber in den vorhandenen Fällen sind es nicht Worte, sondern Begriffe, die sinnlos sein sollen. Z. B. behaupten Whitehead und Russell (und auch andere), dass $\phi(\phi)$ für jedes ϕ bebedeutungslos sei. Zu sagen, dass diese Behauptung nur das Zeichen $\phi(\phi)$ betrifft, heisst, die Frage zu umgehen; weil es doch erstens ein gedachtes Etwas gibt, das das Zeichen $\phi(\phi)$ den Konventionen gemäss bedeuten mag, und zweitens, weil die Gründe,–nämlich das "Vicious Circle Principle" u. s. w. –, woraus die Behauptung entsteht, mit dem Zeichen gar nichts zu tun haben. Die Sinnlosigkeit scheint doch einen sachlichen Inhalt zu haben, aber welchen, lässt sich nicht sagen.

Zunächst gibt es aber einen Sinn, worin alles Denkbare eine Bedeutung hat, nämlich als Begriff. Hier ist ein Begriff als irgendetwas zu verstehen, das identifiziert und von anderen Dingen unterschieden zu werden vermag.* Dann ist es allerdings unbedingt Unsinn zu sagen, dass etwas nicht als Begriff existiere; denn bevor man einen solchen Satz verstehen konnte, muss man sich die Sache als Begriff schon vorgestellt haben. Sogar die "sinnlosen" Gegenstände sind also Begriffe und haben als solche eine Bedeutung.

Unter den Begriffen kommen aber einige vor, die "in sich widerspruchsvoll" sind. Solche sind das oben erwähnte $F(F)$, die grösste Kardinalzahl, die kleinste undefinierbare Ordinalzahl, das kreisförmige Quadrat u. s. w. Diese Begriffe sollen wegen der Widersprüche bedeutungslos sein. Aber die Widersprüche liegen nicht in den Begriffen selbst, sondern in den Eigenschaften, die man ihnen zuordnen möchte: z. B. führt das oben erwähnte $F(F)$ zu einem Widerspruch nur dann, wenn man behauptet, dass es eine Aussage sei; die grösste Kardinalzahl nur dadurch, dass sie wirklich eine Kardinalzahl sei u. s. w.

Betrachten wir nun die sinnlosen Gegenstände überhaupt. Bezieht sich nicht derselbe Gedanke auch auf sie? Ja, die Sinnlosigkeit dieser Begriffe besteht nur darin, dass es Eigenschaften gibt, die sie nicht besitzen. Und zwar, dürfte ich genauer sagen, dass sie den gewöhnlichen Kategorien nicht

---

* Es ist wohl bemerkenswert, dass ein Begriff nach dieser Definition ein Gegenstand und nicht ein Prozees des Denkens ist. Beispiele von Begriffen sind etwa Bismarck, Göttingen, Tier, Regenschirm, rot, Temperatur, Materie, Substanz, Kausalität, Etwas, Funktion, der König von Frankreich, die grösste Kardinalzahl u. s. w.

§1. *Meaningless concepts.* In ordinary logic certain objects (concepts, things) occur that are usually called meaningless. I now ask, just what does this mean? One may well maintain that a word or a symbol is not yet defined in respect to a language. E.g., the word "Kuh" is meaningless in English, although incidentally meaningful in German. But in the present cases it is not the words but the concepts which are said to be meaningless. E.g., Whitehead and Russell (and others as well) maintain that $\phi(\phi)$ is meaningless for every ϕ. To say that this claim concerns only the symbol $\phi(\phi)$ is to evade the question; first because there is yet some object of thought that the symbol $\phi(\phi)$ can mean according to convention, and second because the basis,– namely the "Vicious Circle Principle" etc.–, for the claim has nothing to do with the symbols. The meaninglessness seems to have an objective content, but what exactly is not clear.

Above all there is a sense in which anything thinkable has a meaning, namely as a concept. Here a concept is to be understood as anything that can be identified or differentiated from other things.* Then it is absolute nonsense to say that something does not exist as a concept; for before one can understand such a sentence one must have already envisaged the thing as a concept. Therefore even the "meaningless" objects are concepts and as such have a meaning.

But among the concepts are some which are "contradictory in themselves." Such are the above mentioned $F(F)$, the greatest cardinal number, the least undefinable ordinal number, the round square, etc. These concepts are considered meaningless because of the contradictions. But the contradictions do not lie in the concepts themseleves, but in the properties associated with them–e.g. the above mentioned $F(F)$ leads to a contradiction only if one maintains that it is a proposition; the greatest cardinal number only if it really is a cardinal number, etc.

Now let us consider meaningless objects in general. Do not the same thoughts apply to them also? Yes, the meaninglessness of these concepts consists only in the existence of properties which do not hold of them. And, more exactly, I may certainly say that they

---

*It is well worth remarking that a concept by this definition is an object and not a process of thought. Some examples of concepts are Bismarck, Göttingen, animal, umbrella, red, temperature, matter, substance, causality, entity (*Etwas*), function, the King of France, the greatest cardinal number, etc.

angehören. Diese Kategorien sind in der Tat als etwa inhaltliche Grundbegriffe vorausgesetzt, und nichts wird betrachtet ausser was dazu gehört. Natürlich müssen Begriffe, die mit dem Wesen dieser Kategorien unverträgliche Eigenschaften besitzen, als "sinnlos" aus der Theorie ausgeschlossen werden. Aber gerade in diesem Ausschliessen besteht ein Mangel. Die Aufgabe der Logik ist die Erklärung des Denkens; wenn es von der Erklärung ausgeschlossenes Denken gibt, so ist sie fehlerhaft. Weiterhin sind es genau diese sinnlosen Begriffe, die zu Widersprüchen führen; wenm man sie ausschliesst, so kann man wohl die Antinomien vermeiden, aber nie erklären. Dass etwas ein Begriff ist, ist das einzige Erfordernis, damit man das Ding in der Logik behandeln könne.

Die Kategorie Begriff–oder, wie ich sie nachher nennen werde, um gewisse Nebenbedeutungen zu vermeiden, Etwas,–ist daher die grundlegende Kategorie der Logik überhaupt. Diese Kategorie ist ein Begriff, und seine blosse Betrachtung ist doch in sich widerspruchsfrei. Aus diesen Gründen habe ich sie als Grundkategorie der Theorie vorausgesetzt. Daraus folg eine wichtige Konsequenz: ich brauche mich nicht mehr, mindestens insofern als es die Einführung neuer Gegenstände in die Theorie betrifft, um die Definitionsbereiche von Funktionen zu kümmern. Wenn ich z. B. den Schönfinkelschen Funktionsbegriff als Grundverknüpfung –und Schöfinkel hat schon gezeigt, dass dies allein notwendig ist–, voraussetze, so ist die Zusammensetzung von irgend zwei Begriffen, etwa f und x zu fx, wieder ein Begriff. Z. B. ist die Zusammensetzung des Königs von Frankreich mit der Aussage "der Mond ist aus grünem Käse aufgebaut," ein Begriff, weil sie identifiziert und von anderen Dingen unterschieden zu werden vermag. Unter den so hergestellten Begriffen werden einige "sinnvoll," andere "sinnlos" sein; die Hauptaufgabe der kombinatorischen Logik ist wohl, diese zwei Arten zu unterscheiden.*

§3. *Der Vorrang des Aussagenkalküls.* In den heutigen logischen Theorien bildet der sog. Aussagenkalkül den Grundbestandteil. Ich möchte hier die vielleicht triviale Bemerkumg darlegen, dass dieser Vorgang nicht notwendig ist. Natürlich beginnt die Logik mit Sätzen, die wir fähig sein müssen, zu verstehen. Aber daraus folgt ebensowenig, dass man mit der allgemeinen Theorie von Aussagen anfangen muss, als aus der Tatsache, dass wir es im Aussagenkalkül mit gewissen Aussagefunktionen zu tun haben, folgt, dass wir mit Aussagefunktionen beginnen müssen. Weiterhin sind

---

* Dabei wird die erste Art von Eigenschaften die ich in I A erwähnt habe, auf die Dritte zurückgefürt.

do not belong to the usual categories. These categories are in fact presupposed as somewhat contensive primitive concepts, and nothing is considered which does not belong to them. Naturally, concepts that have properties incompatible with the essence of these categories must be excluded from the theory as "meaningless." But just in this exclusion stands a flaw. The objective of logic is the explanation of thought; if there are thoughts excluded from the explanation, then it is deficient. Furthermore, it is exactly these meaningless concepts which lead to contradictions; if one excludes them, one can avoid the paradoxes but not explain them. That something is a concept is the only requisite for its being subject to treatment in logic.

The category of concept[6]–or, as I shall call it from here on in order to avoid certain extraneous meanings, entity (*Etwas*),[7]–is therefore in general the fundamental category of logic. This category is a concept, and its mere consideration is in itself consistent. For this reason I have postulated it as a fundamental category of the theory. From this follows an important consequence: I need no longer worry, at least insofar as it concerns the introduction of new objects into the theory, about the domains of the functions. E.g., if I postulate Schönfinkel's function concept as the fundamental mode of combination–and Schönfinkel has already shown that this alone is necessary–, then the combination of any two concepts, such as f and x into fx, is again a concept. E.g. the combination of the King of France with the proposition "the moon is made out of green cheese" is a concept, because it can be identified and differentiated from other things. Among the concepts given here, some will be "meaningful" and others "meaningless": the main aim of combinatory logic is to distinguish these two kinds.*

§2. *The priority of the propositional calculus.* In today's logical theories, the so-called propositional calculus constitutes the fundamental component (*Grundbestandteil*). I shall explain here the perhaps trivial remark that this priority is not necessary. Naturally logic begins with theorems that we must be capable of understanding. But it no more follows from that that one must begin with the general theory of propositions than it follows from the fact that we have to do in the propositional calculus with certain propositional functions that we must start with propositional functions. Furthermore, the

---

*Thus the first kind of property that I have listed in I A is reduced to the third.

die Begriffe "Aussage" und "Behauptung" verschieden; dieser ist mit der Wahrheit eng verbunden, jener eine Kategorie von bloss betrachteten Etwasen. Was wir im Anfang verstehen müssen, ist erstens, dass die Sätze Behauptungen sind, und zweitens, dass sie sich vermöge der Regeln zu neuen Behauptungen umgestalten lassen. Was wir mit solchen einfachen Anfängen machen können, zeigt die hier gegebene Theorie an.

Der Begriff von Behauptung muss natürlich als Grundbegriff angenommen werden und zwar als ein solcher, der in der formalen Theorie einem nicht formalen Begriffe entspricht. Diesen nicht formalen Begriff habe ich "*Formel*" genannt.

§3. *Unbeschränkte Universalen.* In Nummer 1 habe ich behauptet, dass der Begriff "Etwas" als grundlegender Begriff angenommen werden darf. Ich behaupte nun ferner, dass es Eigenschaften gibt, die für jedes Etwas gelten. Gegen die Möglichkeit solcher Behauptungen steht aber das wohlbekannte Verbot der nicht prädikativen Begriffsbildungen. Dieses Verbot hat im allgemeinen einen etwa pragmatischen Charakter, und lässt sich folglich nur dadurch widerlegen, dass man wirklich eine dem Verbot widersprechende Theorie aufbauen kann. In den "Principia Mathematica" wird aber dieses Prinzip in der Form des "Vicious Circle Principle," a priori verteidigt. Das dortige Argument ist etwa dies: man könnte nichts über alle Aussagen behaupten, weil dadurch neue Aussagen erschaffen würden, und also es keine bestimmte Gesamtheit von Aussagen gäbe, die den Wirkungsbereich der behaupteten Eigenschaft bildete. Ebensogut aber ist dieses Argument: man könnte nichts über alle Apfelsinen behaupten, weil jedes Jahr neue Apfelsinen erschaffen werden, und es also keine bestimmte Gesamtheit von Apfelsinen gäbe u. s. w. u. s. w. In der Tat begreifen wir allgemeine Urteile nicht in extenso, sondern in intenso–es gibt etwas im Wesen der betreffenden Eigenschaft, wovon wir schliessen können, dass sie für jedes Etwas gilt. Dieser Schluss setzt doch keine Gesamtheit voraus. Es fehlt daher diesem Einwand gegen die Möglichkeit unbeschränkter Universalen ein theoretischer Grund. Infolgedessen mache ich die einfachere und naturgemässere Voraussetzung, dass es solche Universalen gibt; insbesondere, dass solche Eigenschaften wie das reflexive Gesetz der Gleichheit für jedes Etwas gelten.

Wenn wir aber diese Annahme machen, so werden die Grundregeln der Theorie erheblich einfacher. Denn diese Regeln sind Universalen; wenn wir sie als unbeschränkte Universalen auffassen können, so brauchen wir keine Unterscheidungen zwischen verschiedenen Arten von Begriffen inbezug auf die Regeln zu machen. Eine ähnliche Vereinfachung betrifft das ganze Grundgerüst.

concepts "proposition" and "assertion" are different; the latter is closely bound with the truth, the former is a category of entities merely examined. What we must understand at the outset is first, that the theorems are assertions, and second, that they may be transformed by the rules into new assertions. The theory given here will show what we can make with such simple beginnings.

The concept of assertion must naturally be assumed as a primitive concept and moreover as one that expresses in the formal theory a non-formal concept. This non-formal concept I call *"formula"* (*Formel*).

§3. *Absolute universals.* In subsection 1 I asserted that the concept "entity" may be assumed as a primitive concept. I now assert further that there are properties that hold for every entity. Against the possibility of such an assertion stands the well known prohibition on non-predicative concept formation. This prohibition has in general a somewhat pragmatic character, and thus can only be refuted by the actual construction of a theory contradicting the prohibition. In the "Principia Mathematica" this principle is defended a priori in the form of the "Vicious Circle Principle." The argument given there is essentially this: one can assert nothing about all propositions because this would create a new proposition, and so there is no determined collection of propositions that forms the range of the asserted property. One might just as well argue as follows: One can assert nothing about all oranges because every year new oranges are created, and so there is no determined collection of oranges etc. etc. In fact we understand general sentences (*Urteile*) not in extension but in intension–there is something in the essence of the property concerned by which we can conclude that it holds for every entity. This conclusion presupposes no collection. Thus this objection against the possibility of absolute universals lacks a theoretical foundation. Consequently I shall make the simpler and more natural presupposition that there are such universals; in particular, that such properties as the reflexive law of equality hold for every entity.

But when we make this assumption, the primitive rules of the theory become considerably simpler. For these rules are universals; if we can comprehend them as absolute universals we need make no distinction between different kinds of concepts with regard to the rules. A similar simplification concerns the entire primitive frame.

§4. *Theorie und Metatheorie.* Die letzte Betrachtung hat die Methode der Untersuchung zum Gegenstand. Weil die im Anfang vorausgesetzten Regeln sehr einfach sind, ist eine sehr lange Darlegung erforderlich, um diese Entwicklung ausführlich zu vervollständigen. Wir interessieren uns aber nur für die Möglichkeit einer solchen Darlegung. Ich gebe demgemäss, statt die Formeln eine nach der anderen auszuschreiben, hier eine Reihe von Sätzen an, worin ich zeige, dass bestimmte Arten von Etwasen immer Formeln sind. Diese Sätze gehören eigentlich zu der Metamathematik im Hilbertschen Sinne, die mit dieser Mathematik verbunden ist. In den Beweisen dieser Sätze benutze ich nur die einfachen direkten Beweismethoden, und zwar werden die Beweise so aufgestellt, dass sie selbst, natürlich mit Benutzung der Nachweisungen, einen Prozess ergeben, wodurch der Beweis irgendeines in sie eingeschlossenen Falles ausführlich Formel nach Formel auseinandergesetzt werden kann. Unter die Methoden kommen allerdings die vollständige Induktion und andere Eigenschaften der ganzen Zahlen vor, aber dies bedeutet nicht mehr, als dass wir die ganzen Zahlen als Zeichen benutzen, und dass für einige Etwasen, die mit höheren Zahlen bezeichnet werden, jene Beweisprozesse wiederholt werden sollen. Für alle besonderen Fälle hab man eine endliche direkte Schlussfolge.

C. Das Grundgerüst.

Ich gehe jetzt zur formalen Theorie über. Die erste Aufgabe ist natürlich die Darlegung des Grundgerüsts und die Erklärung der in der Ausführung benutzten Festsetzungen.

§I. *Vorbereitende Erklärungen.* Die vorliegende Theorie wird vom abstrakten Gesichtspunkte aus durchgeführt; d. h. als eine Lehre über abstrakte Begriffe, die durch das Grundgerüst selbst definiert werden.* Es ist natürlich wichtig darin die formalen Entwicklumgen selbst und die inhaltlichen Ueberlegungen darüber zu unterscheiden; aber ausserdem habe ich noch die Unterscheidung zwischen der Theorie selbst und ihrer Symbolik festgehalten. Jedoch kann der Leser, der einen anderen Gesichtspunkt vorzieht, wohl die Theorie durch kleine Aenderungen nach seinem eigenen Gesichtspunkt umgestalten. Z. B. ist das, was ich oben über die Endlichkeit der Anzahl von Axiomen ausgesagt habe, mit dem Gesichtspunkt verbunden, dass eine Definition nichts wesentlich Neues in die Theorie hereinbringt, und also von

---

* Ueber das Wesen einer abstrakten Theorie, vgl. meine oben zitierte Abhandlung.

§4. *Theory and metatheory.* The last remark has for its object the method of inquiry. Because the rules postulated at the beginning are very simple, a very long exposition is necessary in order to fully complete this development. But we are interested only in the possibility of such an exposition. I therefore give here, instead of formulas written one after the other, a sequence of theorems in which I show that certain kinds of entities are always formulas. These theorems really belong to the metamathematics in Hilbert's sense, which is connected with this mathematics. In the proofs of these theorems I use only simple, direct methods of proof, and in fact the proofs will be so stated that they themselves, naturally with the use of other proofs, yield a process by which any given case of the proof can be carried out in detail formula by formula, one after the other. Among these methods, of course, are complete induction and other properties of the whole numbers, but this means nothing more than that we use the whole numbers as symbols, and that for some entities, which we label with higher numbers, that the proof process will be repeated. For all special cases one has a finite, direct deduction sequence.

C. The Primitive Frame

I now turn to the formal theory. The first task is naturally the exposition of the primitive frame and the explanation of the conventions used in its realization.

§1. *Preliminary explanations.* The present theory will be presented from an abstract standpoint; i.e., as a theory about abstract concepts that will be defined by the primitive frame itself.* It is naturally important to distinguish between the formal development itself and the contensive considerations about it; but above all I have held fast to the difference between the theory and its symbolic representation. Nevertheless the reader who prefers another standpoint can transform the theory with small modifications to his own standpoint. E.g. what I have asserted above about the finiteness of the number of axioms is connected with the viewpoint that a definition brings nothing really new into the theory and so is different

---

*On the essence of an abstract theory, cf. my paper cited above.

einem Axiom verschieden ist; aber sonst kann man immer einen Satz desselben Inhalts mit etwas anderen Worten aussprechen.

In den Ausführungen brauche ich zwei Arten von Zeichen: erstens die *formalen Zeichen*, welche die Bestandteile der formalen Theorie selbst bedeuten; zweitens die *inhaltlichen Zeichen*, welche eine Bedeutung inbezug auf die inhaltlichen Überlegungen haben. Mit diesen zwei Zeichenarten sind zwei Arten von Definitionen zu betrachten; um diese zu unterscheiden, beschränke ich nunmehr das Wort *Definition* auf die Erklärungen der formalen Zeichen, während die Erklärungen der inhaltlichen Zeichen (bzw. Wörter, die in einem technischen Sinn gebraucht werden), *Festsetzungen* heissen sollen. Die Definitionen und Festsetzungen, die für die ganze Abhandlung beibehalten werden sollen, werden hervorgehoben und als solche ausgezeichnet; neben ihnen mache ich in den einzelnen Beweisen und Erläuterungen Gebrauch von solchen, die nur für den betreffenden Zusammenhang gelten sollen.

Die Definitionen werden weiterhin nun folgendermassen festgestellt. Erstens werden die Zeichen, die in der Darstellung des Grundgerüsts selbst mit gewissen Bestandteilen verbunden sind, dabei definiert. Jede spätere Definition ist dann eine Bestimmung, dass ein nicht schon in demselben Kontext definiertes Zeichen dasselbe wie ein anderes schon definiertes bedeuten soll. Die Identitätsrelation, die in diesem "dasselbe" schon vorliegt, kann, wenn man will, formalisiert werden. Definitionen dürfen sich natürlich nicht nur auf Zeichen im engeren Sinne, sondern auch auf Zeichenverknüpfungen beziehen (vgl. unten Def. 1 und 2).

In der folgenden Darlegung des Grundgerüsts gebe ich neben den zur abstrakten Theorie selbst erforderlichen Eigenschaften auch Andeutungen über die logische Interpretation des betreffenden Begriffs, Regel u. s. w. an. Diese Andeutungen sind aber in Klammern gesetzt.

§2. *Die nicht-formalen Grundbegriffe.* Die folgenden Begriffe seien vorausgesetzt:

a. *Etwas*, eine Kategorie (Ausdeutung, der oben in B I diskutierte Begriff Etwas). Etwase werden durch Buchstaben oder ähnliche Zeichen bezeichnet.

b. *Formel*, eine Kategorie (Ausdeutung: Behauptung).

c. *Anwendung*, eine dyadische Verknüpfung, d. h. eine Zuordnung, wodurch zu jedem geordneten Paar von Etwasen ein eindeutig bestimmtes Drittes zugeordnet wird. Wenn die zwei Etwase durch X bzw. Y bezeichnet sind, dann wird das zugeordnete Dritte die

from an axiom; but otherwise one can always assert a theorem of the same content with different words.

In the development I use two kinds of symbols: first the *formal symbols*, which denote the constituents of the formal theory itself; second the *contensive symbols*, which have a meaning in connection with the contensive considerations. With these two kinds of symbols there are two kinds of definitions to be considered; to distinguish between these, I shall limit from here on the word *definition* to the explanations of the formal symbols, while the explanations of the contensive symbols (or words used in a technical sense) will be called *conventions*. The definitions and conventions which will be retained for the entire work will be set off and distinguished as such; besides them I shall in particular proofs and explanations make use of some which will hold only for the particular context.

In the future definitions will be identified as follows. First the symbols which are connected with certain constituents in the presentation of the primitive frame itself will be thereby defined. Every later definition is then a determination that a symbol not already defined in the same context shall mean the same as another already defined. The identity relation which already exists in this "the same", can be formalized if one wishes. Definitions may be naturally applied not only to symbols in the narrow sense, but also to combinations of symbols (cf. below Defs. 1 and 2).

In the following exposition of the primitive frame I give, besides the properties required for the abstract theory itself, also indications of the logical interpretation of the concepts, rules, etc. in question. But these indications will be set in parentheses.

§2. *The non-formal primitive ideas.* The following concepts will be postulated:

a. *Entity* (*Etwas*), a category (Interpretation, the concept of entity discussed above in B 1). Entities will be denoted by letters or similar symbols.

b. *Formula* (*Formel*), a category (Interpretation: assertion).

c. *Application* (*Anwendung*), a dyadic mode of combination, i.e. a relation (*Zuordnung*) by which to every ordered pair of entities a uniquely determined third entity is associated. If the two entities are denoted by X and Y, then the associated third entity is called the application

Anwendung von X auf Y genannt und durch (XY) bezeichnet. (Ausdeutung: der Schönfinkelsche Funktionsbegriff, erweitert wie in B1 oben).

§3. *Die formalen Grundfbegriffe.* Von diesen Begriffen wird nichts vorausgesetzt, ausser dass sie Etwase sind. Ich schreibe also hier nur ihre Zeichen und Ausdeutungen:

B (Die Schönfinkelsche Zusammensetzungsfunktion).
C (Die Schönfinkelsche Vertauschungsfunktion).
W (eine Funktion derselben Art wie die vorhergehende, durch die Regel W [s. unten] definiert. Wir dürften sie die Verdoppelungsfunktion nennen).
K (Die Schönfinkelsche Konstanzfunktion).
Q (Gleichheit, d. h. logische, nicht, symbolische Identität).
Π (Allzeichen).
P (Implikation).
Λ (Konjunktion –das Und– Vgl. unten, Reg. Λ).

Die vorhergehenden Begriffe sind die einzigen, die in dieser Abhandlung erwähnt sind; für die weitere Durchführung der Logik brauchen wir natürlich gewisse andere, z. B. Aussage, Negation, Funktion, Seinzeichen (wenn nicht definiert) u. s. w.

§4. *Symbolische Festsetzungen.*

Festsetzung 1. Wenn X ein Etwas ist, so bezeichne ich mit $\vdash X$ den Satz, dass X eine Formel ist, und zwar sowohl wenn dieser Satz behauptet, festgesetzt oder bloss betrachtet ist. Weiterhin dürfen dann die äusseren Klammern von X wenn es solche gibt, wegbleiben.

Festsetzung 2. Wenn X und Y Etwase sind, so bezeichne ich mit $X \equiv Y$ den Satz, dass X und Y dasselbe Etwas sind, d. h. dass X und Y als Zeichen betrachtet, dasselbe Etwas bedeuten. Unter diesen Umständen werde ich auch manchmal sagen, dass X und Y *identisch* sind. Die äusseren Klammern von X und Y dürfen auch hier wegbleiben.

Def. 1. Immer wenn $X_1, X_2 \cdots, X_n$ Etwase sind, gilt

$$(X_1 X_2 X_3 \cdots X_n) \equiv (\cdots(((X_1 X_2)X_3)\cdots)X_n).$$

Def. 2. Immer wenn X und Y Etwase sind, gilt

$$(X = Y) \equiv (QXY).$$

of X to Y and is denoted (XY). (Interpretation: the Schönfinkel function concept, extended as in B 1 above).

§3. *The formal primitive ideas.* Nothing will be postulated about these concepts except that they are entities. I thus write here only their symbols and interpretation:

- B (The Schönfinkel composition function).
- C (The Schönfinkel interchange function).
- W (A function of the same kind as the foregoing, defined by Rule W [see below]. We shall call this the doubling function).
- K (The Schönfinkel constancy function).
- Q (Equality, i.e. logical, not symbolic identity).
- Π (Universal quantifier).
- P (Implication).
- Λ (Conjunction–the and–cf. Rule Λ below).

The foregoing concepts are the only ones which are mentioned in this work; for the further development of logic we shall naturally use certain others, e.g. proposition, negation, function, existential quantifier (if not defined), etc.

§4. *Symbolic Conventions.*

Convention 1 If X is an entity, I denote by $\vdash X$ the theorem that X is a formula, and in fact I use this same notation whether this theorem is asserted, postulated, or only considered. Furthermore, the outer parentheses of X, if any, may be omitted.

Convention 2. If X and Y are entities, I denote by $X \equiv Y$ the theorem that X and Y are the same entity, i.e. that X and Y considered as signs mean the same entity. Under these circumstances I shall also often say that X and Y are *identical*. The outer parentheses of X and Y may also be omitted here.

Def. 1. Whenever $X_1, X_3, \cdots, X_n$ are entities,

$$(X_1 X_2 X_3 \cdots X_n) \equiv (\cdots (((X_1 X_2) X_3) \cdots) X_n).$$

Def. 2. Whenever X and Y are entities,

$$(X = Y) \equiv (QXY).$$

Furthermore,

Links dürfen weiter die äusseren Klammern von X und Y wegbleiben, solange nur X und Y nicht selbst von der Form $(U = V)$ sind.

Def. 3. $I \equiv (WK)$.

Festsetzung 3. Die Relation zwischen zwei Etwasen X und Y, welche durch $\vdash X = Y$ ausgedrückt wird, heisst *Gleichheit*, und unter diesen Umständen sage ich, dass X und Y gleich sind.

§5. *Axiome.* Die hier gegebenen Axiome reichen nur für diese Abhandlung aus. Die kombinatorischen Axiome werden unten (s. II B 5)* in viel übersichtlicherer Form wiedergegeben.

a. *Axiom der Identität.*

   Ax. Q. $\vdash \Pi(W(CQ))$ (Reflexives Gesetz, s. unten, Abschn. D).

b. *Kombinatorische Axiome.*

   Ax. B. $\vdash C(BB(BBB))B = B(BB)B$.
   Ax. C. $\vdash C(BB(BBB))C = B(BC)(BBB)$.
   Ax. W. $\vdash C(BBB)W = B(BW)(BBB)$.
   Ax. K. $\vdash C(BBB)K = B(BK)I$.
   Ax. I_1. $\vdash CBI = B(BI)I$.†
   Ax. (BC). $\vdash BBC = B(B(BC)C)(BB)$.
   Ax. (BW). $\vdash BBW = B(B(B(B(BW)W)(BC))B(BB))B$.
   Ax. (BK). $\vdash BBK = BKK$.
   Ax. $(CC)_1$. $\vdash BCC = B(BI)$.
   Ax. $(CC)_2$. $\vdash B(B(BC)C)(BC) = B(BC(BC))C$.
   Ax. (CW). $\vdash BCW = B(B(BW)C)(BC)$.
   Ax. (CK). $\vdash BCK = BK$.
   Ax. (WC). $\vdash BWC = W$.
   Ax. (WW). $\vdash BWW = BW(BW)$.
   Ax. (WK). $\vdash BWK = BI$.
   Ax. I_2. $\vdash BI = I$.

§6. *Regeln.* Die folgenden Regeln sind vermutlich für die gesamte Mathematik und Logik hinreichend. Die zwei letzteren werden nur in einem späteren Stadium benutzt, aber sie sind hier der Vollständigkeit halber gegeben.

---

* s. d. für Bemerkungen über die Bedeutung dieser Axiome.

† Dieser Axiom kann aus den anderen bewiesen werden. s. unten. II D 2, Satz 6.

the outer parentheses of X and Y on the left may be omitted provided only that X and Y are not themselves of the form $(U = V)$.

Def. 3. $I \equiv (WK)$.

Convention 3. The relation between two entities X and Y which will be written $\vdash X = Y$, will be called *equality*, and in these circumstances I shall say that X and Y are *equal*.

§5. *Axioms.* The axioms given here suffice only for this work. The combinatory axioms will be given again below (see II B 5)* in a much clearer form.

a. *Axiom of identity.*

Ax. Q. $\vdash \Pi(W(CQ))$ (Reflexive law, see below, section D).

b. *Combinatory axioms.*

Ax. B.	$\vdash C(BB(BBB))B = B(BB)B$.
Ax. C.	$\vdash C(BB(BBB))C = B(BC)(BBB)$.
Ax. W.	$\vdash C(BBB)W = B(BW)(BBB)$.
Ax. K.	$\vdash C(BBB)K = B(BK)I$.
Ax. I_1.	$\vdash CBI = B(BI)I$.*
Ax. (BC).	$\vdash BBC = B(B(BC)C)(BB)$.
Ax. (BW).	$\vdash BBW = B(B(B(BW)W)(BC))(B(BB))B$.
Ax. (BK).	$\vdash BBK = BKK$.
Ax. $(CC)_1$.	$\vdash BCC = B(BI)$.
Ax. $(CC)_2$.	$\vdash B(B(BC)C)(BC) = B(BC(BC))C$.
Ax. (CW).	$\vdash BCW = B(B(BW)C)(BC)$.
Ax. (CK).	$\vdash BCK = BK$.
Ax. (WC).	$\vdash BWC = W$.
Ax. (WW).	$\vdash BWW = BW(BW)$.
Ax. (WK).	$\vdash BWK = BI$.
Ax. I_2.	$\vdash BI = I$.

§6. *Rules.* The following rules are conjectured to be sufficient for the whole of mathematics and logic. The last two will only be used in a later study, but they are given here for completeness.

---

*See there for remarks on the meaning of these axioms.
†This axiom can be proved from the others. See below II D 2, Theorem 6.

Reg. *E*. Wenn X und Y Etwase sind, dann ist immer XY ein Etwas. (Ueber die Ausdeutung dieser Regel vgl. oben, B 1. Sie ist so einfach, dass sie nie wieder erwähnt wird, sondern nur implizit auftritt).

Reg. Q_1. Wenn X und Y Etwase sind, und 1) $\vdash X$, 2) $\vdash QXY$ (bzw. $\vdash X = Y$), dann $\vdash Y$.

Reg. Q_2. Wenn X, Y, Z Etwase sind, und $\vdash QXY$ (bzw. $\vdash X = Y$), dann $\vdash Q(ZX)(ZY)$ (d. h. $\vdash ZX = ZY$).

Reg. Π. Wenn X und Y Etwase sind, und $\vdash \Pi X$, dann $\vdash XY$. (Die Ausdeutumg dieser Regel ist das sog. Prinzip von Aristoteles).

Reg. *B*. Wenn X, Y, Z Etwase sind, dann $\vdash BXYZ = X(YZ)$.

Reg. *C*. Wenn X, Y, Z Etwase sind, dann $\vdash CXYZ = XZY$.

Reg. *W*. Wenn X, Y Etwase sind, dann $\vdash WXY = XYY$.

Reg. *K*. Wenn X, Y Etwase sind, dann $\vdash KXY = X$.

Reg. *P*. Wenn X, Y Etwase sind, sodass 1) $\vdash X$, 2) $\vdash PXY$, dann $\vdash Y$. (Dies ist die wohlbekannte Schlussregel).

Reg. Λ. Wenn X, Y Etwase sind, sodass 1) $\vdash X$, 2) $\vdash Y$, dann $\vdash \Lambda XY$. (Diese Regel kann natürlich vermieden werden, wenn in den formalen Entwicklungen der Theorie geschlossen werden kann, dass für beliebige X, Y

$$\vdash PX(PY(\Lambda XY),$$

d. h. in gewöhnlicherer Schreibweise,

$$X \to (Y \to \langle X \ \& \ Y \rangle) \ .$$

Aber diese Aussage ist mit der logischen Bedeutung der Implikation unverträglich. Wenn wir diese logische Implikation eintreten lassen wollen, so müssen wir P und Λ als unabhängige Grundbegriffe, und die beiden Regeln P und Λ voraussetzen. [Vgl. Lewis, C. J.–*A Survey of Symbolic Logic*. Berkeley 1918, Chap. V]).

D. Die Eigenschaften der Gleichheit.

Satz 1. *Wenn X ein Etwas ist, dann* $\vdash QXX$.
Beweis:

	$\vdash \Pi(W(CQ))$	(Ax. Q).
$\therefore$	$\vdash W(CQ)X$	(Reg. Π, Hp.).
Aber	$\vdash Q(W(CQ)X)(CQXX)$	(Reg. W).
$\therefore$	$\vdash CQXX$	(Reg. Q_1).
Aber	$\vdash Q(CQXX)(QXX)$	(Reg. C).
$\therefore$	$\vdash QXX$ w. z. b. w.	(Reg. Q_1).

Rule E. If X and Y are entities, then (XY) is always an entity. (On the interpretation of this rule, cf. above, B 1. It is so simple that it will never be mentioned further, but will only appear implicitly.)

Rule Q_1. If X and Y are entities, and 1) $\vdash X$, 2) $\vdash QXY$ (or $\vdash X = Y$), then $\vdash Y$.

Rule Q_2. If X, Y, Z are entities and $\vdash QXY$ (or $\vdash X = Y$), then $\vdash Q(ZX)(ZY)$ (i.e., $\vdash ZX = ZY$).

Rule Π. If X and Y are entities and $\vdash \Pi X$, then $\vdash XY$. (The interpretation of this rule is the so-called Principle of Aristotle.)

Rule B. If X, Y, Z are entities, then $\vdash BXYZ = X(YZ)$.

Rule C. If X, Y, Z are entities, then $\vdash CXYZ = XZY$.

Rule W. If X, Y are entities, then $\vdash WXY = XYY$.

Rule K. If X, Y are entities, then $\vdash KXY = X$.

Rule P. If X, Y are entities so that 1) $\vdash X$, 2) $\vdash PXY$, then $\vdash Y$. (This is the well known rule of inference).[8]

Rule Λ. If X and Y are entities so that 1) $\vdash X$, 2) $\vdash Y$, then $\vdash \Lambda XY$. (This rule can naturally be omitted if we can conclude in the development of the formal theory that for every X, Y
$$\vdash PX(PY(\Lambda XY)),$$
i.e. in more usual notation $X \to (Y \to (X \& Y))$.
But this proposition is incompatible with the logical meaning of implication. If we want to let this logical implication occur, we must take P and Λ as independent postulates and postulate both rules P and Λ. [Cf. Lewis, C. I.–*A Survey of Symbolic Logic.* Berkeley, 1918, Chap. V]).

D. The Properties of Equality.

Theorem 1. *If X is an entity, then $\vdash QXX$.*

Proof.

	$\vdash \Pi(W(CQ))$	(Ax. Q).
∴	$\vdash W(CQ)X$	(Rule Π, Hp.).
But	$\vdash Q(W(CQ)X)(CQXX)$	(Rule W).
∴	$\vdash CQXX$	(Rule Q_1).
But	$\vdash Q(CQXX)(QXX)$	(Rule C).
∴	$\vdash QXX$ qed.	(Rule Q_1).

SATZ 2. *Wenn X und Y Etwase sind und $\vdash QXY$, dann $\vdash QYX$*

Beweis :
$\quad\vdash QXY$ (Hp.)
$\therefore\quad \vdash Q(CQXX)(CQXX)$ (Reg. Q_2, $Z \equiv CQX$).
Aber $\quad \vdash CQXX$ (vorvorletzte Formel im Beweis von Satz 1).
$\therefore\quad \vdash CQXY$ (Reg. Q).
$\therefore\quad \vdash QYX$ w. z. b. w. (Reg. C und Q_1).

SATZ 3. *Wenn X, Y, Z Etwase sind, und 1) $\vdash QXY$ 2) $\vdash QYZ$, dann $\vdash QXZ$.*

Beweis:
$\quad \vdash QYZ$ (Hp. 2).
$\quad \vdash Q(QXY)(QXZ)$ (Reg. Q_2).
$\therefore\quad \vdash QXZ$ w. z. b. w. (Hp. 1 und Reg. Q_1).

SATZ 4. *Wenn X, Y, Z Etwase sind, und $\vdash QXY$, dann $\vdash Q(XZ)(YZ)$.*

Beweis:
$\vdash Q(C(B(WK))ZX)(C(B(WK))ZY)$ (Hp. und Reg. Q_2).
$\vdash Q(C(B(WK))ZX)(B(WK)XZ)$ (Reg. C).
$\vdash Q(B(WK)XZ)(WK(XZ))$ (Reg. B).
$\vdash Q(WK(XZ))(K(XZ)(XZ))$ (Reg. W).
$\vdash Q(K(XZ)(XZ))(XZ)$ (Reg. K).

aus den letzten vier Formeln und Satz 3, $\vdash Q(C(B(WK))ZX)(XZ)$. In ähnlicher Weise folgt dass $\vdash Q(C(B(WK)ZY)(YZ)$.

Aus der ersten und den zwei letzten Formeln und den Sätzen 2 und 3. $\vdash Q(XZ)(YZ)$. w. z. b. w.

SATZ 5. *Wenn X und Y Etwase sind, und $X \equiv Y$, dann $\vdash QXY$.*

Beweis: Aus Satz 1 $\vdash QXX$.

Weil Y dasselbe wie X bedeutet, so folgt $\vdash QXY$. w. z. b. w.

SATZ 6. *Gleichheit ist eine reflexive, symmetrische und transitive Relation und zwar derart, dass aus $\vdash X = Y$ für ein beliebiges Etwas Z fotgt*

$$\vdash ZX = ZY, und \vdash XZ = YZ,$$

und weiterhin aus $\vdash X$ und $\vdash X = Y$ folgt $\vdash Y$.

Beweis: Dieser Satz ist nur eine Zusammenfassung von Reg. Q_1 und Q_2 und Sätzen 1-4 inbezug auf C 4. Def. 2 und der Bedeutung von $\equiv$. Er lässt

Theorem 2. *If X and Y are entities, and $\vdash QXY$, then QYX.*

Proof.

$\qquad\quad \vdash QXY$ (hp.)
$\therefore\quad \vdash Q(CQXX)(CQXY)$ (Rule Q_2, $Z \equiv CQX$).
But $\;\vdash CQXX$ (third from last formula in proof of Th. 1).
$\therefore\quad \vdash CQXY$ (Rule Q).
$\therefore\quad \vdash QYX$ qed. (Rules C and Q_1).

Theorem 3. *If X, Y, Z are entities, and 1) $\vdash QXY$, 2) $\vdash QYZ$, then $\vdash QXZ$.*

Proof.

$\qquad \vdash QYZ$ (Hp. 2).
$\qquad \vdash Q(QXY)(QXZ)$ (Rule Q_2).
$\therefore\quad \vdash QXZ$ qed. (Hp. 1 and Rule Q_1).

Theorem 4. *If X, Y, Z are entities, and $\vdash QXY$, then $\vdash Q(XZ)(YZ)$.*

Proof.

$\vdash Q(C(B(WK))ZX)(C(B(WK))ZY)$ (Hp. and Rule Q_2).
$\vdash Q(C(B(WK))ZX)(B(WK)XZ)$ (Rule C).
$\vdash Q(B(WK)XZ)(WK(XZ))$ (Rule B).
$\vdash Q(WK(XZ))(K(XZ)(XZ))$ (Rule W).
$\vdash Q(K(XZ)(XZ))(XZ)$ (Rule K).

from the last four formulas and Theorem 3, $\vdash Q(C(B(WK))ZX)(XZ)$. In a similar way it follows that $\vdash Q(C(B(WK))ZY)(YZ)$.

From the first and the last two formulas and Theorems 2 and 3,
$\qquad \vdash Q(XZ)(YZ)$. qed.

Theorem 5. *If X and Y are entities, and $X \equiv Y$, then $\vdash QXY$.*

Proof. By Theorem 1, $\vdash QXX$.
Because Y is the same as X, so it follows that $\vdash QXY$.

Theorem 6. *Equality is a reflexive, symmetric, and transitive relation, and it is also true that if $\vdash X = Y$ for any entity Z, then*
$\qquad \vdash ZX = ZY$, *and* $\vdash XZ = YZ$,
and furthermore from $\vdash X$ and $\vdash X = Y$ it follows that $\vdash Y$.

Proof. This theorem is only a summary of Rules Q_1 and Q_2 and Theorems 1–4 in reference to C4. Def. 2 and the meaning of $\equiv$. It

sich aber auch, abgesehen von der Bedeutung von $\equiv$, nur mit Benutzung von Satz 5 ableiten. Z. B. beweise ich die letzte Behauptung.

Angenommen $\vdash X = Y$,
dann $\vdash QXY$ (C4. Def. 2, Satz 5, Reg. Q_1),
daher aus $\vdash X$,
folgt $\vdash Y$ (Reg. Q_1).

SATZ 7. $\mathfrak{X}$ *sei ein Etwas, das aus gegebenen Etwasen* $X_1, X_2, \cdots, X_n$ *(unter anderen) durch Anwendung entsteht, und* $\mathfrak{Y}$ *sei der Ausdruck, welcher dann entsteht, wenn man* $X_1, X_2, \cdots X_n$ *in* $\mathfrak{X}$ *durch Etwase* Y_1, $Y_2 \cdots Y_n$ *ersetzt. Ferner sei für alle* $i = 1, 2 \cdots n$ *entweder* $\vdash X_i = Y_i$ *oder* $\vdash Y_i = X_i$. *Dann* $\vdash \mathfrak{X} = \mathfrak{Y}$.

Beweis: Wir können annehmen, dass X_i nur eimmal in $\mathfrak{X}$ vorkommt, weil der Fall, dass einige X_i nochmals vorkommen, sich, auf den Fall, dass sie alle nur einmal vorkommen, durch Vergrösserung des n zurückführen lässt (die X_i brauchen nicht alle verschieden zu sein). Weiterhin können wir uns auf den Fall $n = 1$ beschränken, weil es sonst immer eine Reihe von Ausdrücken $\mathfrak{X}_0 \equiv \mathfrak{X}, \mathfrak{X}_1, \mathfrak{X}_2, \cdots \mathfrak{X}_n \equiv \mathfrak{Y}$ derart gibt, dass $\mathfrak{X}_{i+1}$ sich aus $\mathfrak{X}_i$ durch Einsetzung von Y_i statt X_i erzeugt, und wenn wir dann bewiesen haben, dass

$\vdash \mathfrak{X}_i = \mathfrak{X}_{i+1}$
so folgt $\vdash \mathfrak{X}_0 = \mathfrak{X}_n$ (Satz 6),
also $\vdash \mathfrak{X} = \mathfrak{Y}$ (Satz 5, 6).

Infolgedessen nehmen wir an, dass $\mathfrak{Y}$ sich aus $\mathfrak{X}$ durch Einsetzung von Y statt X erzeugt. Dann gibt es zwei Reihen von Etwasen $X_1, X_2 \cdots X_m, Y_1, Y_2, \cdots Y_m$ derart, dass

$X_1 \equiv X,$ $Y_1 \equiv Y,$
$X_{i+1} \equiv X_i Z_i$ oder $X_{i+1} \equiv Z_i X_i,$
und $Y_{i+1} \equiv Y_i Z_i$ bzw. $Y_{i+1} \equiv Z_i Y_i,$
$X_m \equiv \mathfrak{X},$ $Y_m \equiv \mathfrak{Y}.$

Dann haben wir zuerst $\vdash X_1 = Y_1$ (Hp., Satz 5,–vielleicht Satz 3), und weiter aus $\vdash X_i = Y_i$ folgt $\vdash X_{i+1} = Y_{i+1}$ (Satz 6). Daher durch Induktion $\vdash X_m = Y_m$,
daher $\vdash \mathfrak{X} = \mathfrak{Y}$. w. z. b. w.

SATZ 8. *Die zwei Sätze* 1) $\mathfrak{X} \equiv \mathfrak{Y}$ *und* 2) : *"Aus den vorangegebenen Säizen der Form* $X_i \equiv Y_i$ *kann der Satz* $\vdash \mathfrak{X} = \mathfrak{Y}$ *allein mit Benutzung von Satz 5 und die Eigenschaften der Gleichheit bediesen werden," sind äquivalent.*

can also be proved, given the meaning of $\equiv$, using only Theorem 5. E.g., I prove the last assertion.

Given $\vdash X = Y$,
then $\vdash QXY$ (C4. Def. 2, Theorem 5, Rule Q_1),
therefore if $\vdash X$,
then $\vdash Y$ (Rule Q_1).

Theorem 7. *Let $\mathfrak{X}$ be an entity, that is composed out of given entities $X_1, X_2, \cdots, X_n$ (and others) by application, and let $\mathfrak{Y}$ be the expression which results when the entities $X_1, X_2, \cdots, X_n$ are replaced by $Y_1, Y_2, \cdots Y_n$ respectively. Further, for all $i = 1, 2, \cdots, n$ suppose that either $\vdash X_i = Y_i$ or $\vdash Y_i = X_i$. Then $\vdash \mathfrak{X} = \mathfrak{Y}$.*

Proof. We can assume that X_i occurs only once in $\mathfrak{X}$, because the case that X_i occurs more than once can be taken care of under the case that all the X_i occur only once by increasing the value of n (the X_i need not all be different). Furthermore, we can limit ourselves to the case $n = 1$ because otherwise there is a sequence of expressions $\mathfrak{X}_0 \equiv \mathfrak{X}, \mathfrak{X}_1, \mathfrak{X}_2, \cdots, \mathfrak{X}_n$ such that $\mathfrak{X}_{i+1}$ is obtained from $\mathfrak{X}_i$ by replacing X_i by X_{i+1}, and we will then
$$\vdash \mathfrak{X}_i = \mathfrak{X}_{i+1}$$
have proved that and so follows $\vdash \mathfrak{X}_0 = \mathfrak{X}_n$ (Theorem 6),
thus $\vdash \mathfrak{X} = \mathfrak{Y}$ (Theorems 5, 6).

Consequently we assume that $\mathfrak{Y}$ is obtained from $\mathfrak{X}$ by the replacement of X by Y. Then there are two sequences of entities $X_1, X_2, \cdots, X_m$ and $Y_1, Y_2, \cdots, Y_m$ such that

and
$X_1 \equiv X,$ $Y_1 \equiv Y,$
$X_{i+1} \equiv X_i Z_i$ or $X_{i+1} \equiv Z_i X_i,$
$Y_{i+1} \equiv Y_i Z_i$ or $Y_{i+1} \equiv Z_i Y_i,$
$X_m \equiv \mathfrak{X},$ $Y_m \equiv \mathfrak{Y}.$

Then we have first that $\vdash X_1 = Y_1$ (Hp., Theorem 5, –perhaps Theorem 3),

and further, if $X_i = Y_i$, then $\vdash X_{i+1} = Y_{i+1}$ (Theorem 6).

Hence, by induction, $\vdash X_m = Y_m$,

whence $\vdash \mathfrak{X} = \mathfrak{Y}$ qed.

Theorem 8. *The two theorems 1) $\mathfrak{X} \equiv \mathfrak{Y}$ and 2) "From previously given theorems of the form $X_i \equiv Y_i$ the theorem $\vdash \mathfrak{X} = \mathfrak{Y}$ can be proved using only Theorem 5 and properties of equality," are equivalent.*

Beweis: Der allgemeinste Satz der Form $\mathfrak{X} \equiv \mathfrak{Y}$, den wir nach der Bedeutung van $\equiv$ aus den vorangegebenen schliessen können, ist ein Spezialfall von Satz 7 mit Gleichheit durch Identität ersetzt. Aber Satz 7 gibt auch den allgemeinsten Satz, der Form $\vdash \mathfrak{X} = \mathfrak{Y}$ welcher aus $\vdash X_i = Y_i$ durch Benutzung der Eigenschaften der Gleichheit bewiesen werden kann. Damit wird die Behauptung bewiesen.

Festsetzung 1. Unter

$$\vdash X_1 = X_2 = X_3 \cdots = X_n$$

verstehen wir

$\vdash X_1 = X_2$ und $\vdash X_2 = X_3$ und $\cdots$ und $\vdash X_{n=1} = X_n$,

(woraus insbesondere folgt $\vdash X_1 = X_n$. Die Eigenschaften werden der Gleichheit hiernach im allgemeinen nicht besonders erwähnt).

SATZ 9. *Wenn X ein Etwas ist*

$$\vdash IX = X.$$

Beweis: $\vdash IX = WKX$ (C4. Def. 3, Satz 4),
$ = KXX$ (Regel W),
$ = X$ (Regel K).

Bemerkung: Anstatt B, C, W und K hat Schönfinkel K und S als Grundbegriffe gewählt. Dazu gehören die Regeln

$\vdash KXY = X$ $\quad$ für beliebige Etwase X, Y ;
$\vdash SXYZ = XZ(YZ)$ $\quad$ für beliebige Etwase X, Y, Z,

woraus sich B, C, W und I wie folgt definieren lassen:

$$B \equiv S(KS)K, \ C \equiv S(BBS)(KK), \ W \equiv SS(SK), \ I \equiv SKK.$$

In dieser Weise kann man wohl die Anzahl der Grundbegriffe und Regeln vermindern. Der Beweis der Eigenschaften der Gleichheit ist aber etwas schwieriger, und man muss vermutlich den obigen Satz 4 als Grundregel voraussetzen.

Proof. The most general theorem of the form $\mathfrak{X} \equiv \mathfrak{Y}$, that we can obtain from the meaning of $\equiv$ in the foregoing, is a special case of Theorem 7 with equality replaced by identity. But Theorem 7 also gives the most general theorem of the form $\vdash \mathfrak{X} = \mathfrak{Y}$ which can be proved from $\vdash X_i = Y_i$ by the use of properties of equality. In this way the result is proved.

Convention 1. By

$$\vdash X_1 = X_2 = X_3 = \cdots = X_n$$

we understand

$$\vdash X_1 = X_2 \text{ and } \vdash X_2 = X_3 \text{ and } \cdots \text{ and } \vdash X_{n-1} = X_n,$$

(from which follows $\vdash X_1 = X_n$. The properties of equality will in general not be further mentioned.)

Theorem 9. *If X is an entity, then*

$$\vdash IX = X.$$

Proof.
$$\begin{aligned}
\vdash IX &= WKX &&\text{(C4. Def. 3, Theorem 4)}, \\
&= KXX &&\text{(Rule } W\text{)}, \\
&= X &&\text{(Rule } K\text{)}.
\end{aligned}$$

Remark. Instead of B, C, W, and K, Schönfinkel chose K and S as basic concepts. To these belong the rules
$\vdash KXY = X$ for any entities X, Y ;
$\vdash SXYZ = XZ(YZ)$ for any entities X, Y, Z,
from which B, C, W and I can be defined as follows:

$$B \equiv S(KS)K, \ C \equiv S(BBS)(KK), \ W \equiv SS(SK), \ I \equiv SKK.$$

In this way the number of basic concepts and rules can be reduced. But the proof of the properties of equality is somewhat more difficult, and Theorem 4 must then be assumed as a basic rule.

KAPITEL II. DIE LEHRE DER KOMBINATOREN.

A. Einleitung.

Festsetzung 1. Unter einer *Kombination* von Etwasen $X_1, X_2 \cdots X_n$ verstehen wir ein Etwas, das sich aus $X_1 \cdots X_n$ durch Anwendung aufbauen lässt. Genauer definieren wir so:

1) Jeder X_i ($i = 1, 2, \cdots, n$) ist eine Kombination won $X_1 \cdots X_n$.

2) Wenn X und Y Kombinationen von $X_1, \cdots X_n$ sind, so ist auch (XY) eine solche.

Festsetzung 2. Unter einem *Kombinator* verstehen wir eine Kombination von B, C, W und K.

In diesem Kapital untersuchen wir die Kombinatoren als solche. Das Hauptergebnis betrifft aber eine Verwandtschaft mit den Substitutionsprozessen. Es wird gezeigt, dass alle die in der gewöhnlichen Logik durch freie Variablen angedeuteten Verknüpfungen sich durch Kombinatoren deffinieren lassen, und zwar so, dass alle die Eigenschaften, die sie dort haben, auch aus unserem Grundgerüst ableitbar sind.

Um dieses Problem schärfer zu formulieren, betrachten wir diese Operationen und ihre Ausdrucksweise näher. Es soll genau erklärt werden, was sie sind und welche Eigenschaften in Betracht kommen.

Zunächst dienen natürlich die Variablen nur dazu, die Leerstelle in den verschiedenen Funktionen zu unterscheiden. Infolgedessen darf ich für sie eine besondere Bezeichnungsweise fordern, und zwar die folgende: Man beginnt mit gewissen Grundkonstanten und Grundfunktionen. In diesen sollen die Leerstelle mit den Zeichen $x_1, x_2 \cdots x_n$ konsekutiv (d. h. ohne Auslassungen)* numeriert werden, und zwar in einer Anordnung, die am Anfang beliebig zu wählen ist, aber danach festbleiben soll. Dann werden die folgenden Operationen für die Erschaffung neuer Funkionen und Aussagen erlaubt.

---

* [Oder genauer, wenn x_e als Argument vorkommt, so soll auch x_k für $k < e$ vorkommen. (P. Bernays)].

† In der gewöhnlichen Theorie sagt man zuweilen, dass ese zwischen $\phi(x,y)$ und $\phi(y,x)$ keinen Unterschied gebe, weil es nicht darauf ankomme, wie man die Variablen bezeichne. Aber wenn man eine Funktion $\phi(x,y)$ hat, so kann man andere Funktionen $\psi(x,y)$, $\chi(x)$ definieren, sodass

$$\psi(x,y) \equiv_{xy} \phi(y,x) \qquad \chi(x) \equiv_x \phi(x,x).$$

Dann ist der Inhalt meiner Forderung dieser: wenn wir z. B. ϕ mit $\phi(x_1, x_2)$ bezeichnen, dann werden ψ und χ mit $\phi(x_2, x_1)$ bzw. $\phi(x_1, x_1)$ bezeichnet. Natürlich haben wir hier im allgemeinen drei verschicdene Funktionen.

Chapter II. The Theory of Combinators.

A. Introduction

Convention 1. By a *combination* of entities $X_1, X_2, \cdots, X_n$ we understand an entity which can be built up from $X_1, X_2, \cdots, X_n$ by application. More exactly, the definition is as follows:

1) Every $X_i (i = 1, 2, \cdots, n)$ is a combination of $X_1, X_2, \cdots, X_n$.

2) If X and Y are combinations of $X_1, X_2, \cdots, X_n$, then (XY) is also such a combination.

Convention 2. By a *combinator*, we understand a combination of $B, C, W,$ and K.

In this chapter we study the combinators as such. But the main result concerns a relation with the process of substitution. It will be shown that everything (*alle Verknupfungen*) in ordinary logic which can be indicated by free variables can be defined by combinators, and thus that all the properties that they have there can be derived from our primitive frame.

To formulate this problem more sharply, we examine these operations and their modes of expression more closely. It will be explained exactly what they are and what properties are relevant.

First, variables only serve the purpose of differentiating between different empty places in different functions. Consequently, I may require a certain mode of expression for them, which is as follows: one begins with certain basic constants and basic functions. In these the blanks shall be numbered consecutively (i.e., without omission)* by the signs $x_1, x_2, \cdots, x_n$, and in an order which can be freely chosen at the beginning but after that remains fixed. Then the following operations for the creation of new functions and statements will be possible.[†9]

---

*[Or more exactly, if x_e appears as an argument, then every x_k for $k < e$ also appears. (P. Bernays)]

[†]In the usual theory one often says that between $\phi(x,y)$ and $\phi(y,x)$ there is no difference, because it does not matter how one uses variables. But when one has a function $\phi(x,y)$, one can define other functions $\psi(x,y), \chi(x)$ so that
$$\psi(x,y) \equiv_{x,y} \phi(y,x) \qquad \chi(x) \equiv_x \phi(x,x).$$
Then the content of my requirement is this: if we e.g. denote ϕ by $\phi(x_1, x_2)$, then ψ and χ will be denoted by $\phi(x_2, x_1)$ and $\phi(x_1, x_1)$ respectively. Naturally we have here in general three different functions.

1) *Umwandlung* einer Funktion in eine zweite, die sich von der ersten nur darin unterscheidet, dass die Variablen anders numeriert sind. Diese neue Numerierung soll die obigen Bedingungen der Konsekutivität erfüllen, doch ist es erlaubt, Leerstellen, die ursprünglich verschiedene Zeichen trugen, dasselbe Zeichen zu geben, aber nicht umgekehrt. Z. B. aus $\phi(x_1, x_2)$ werden $\phi(x_2, x_1)$ und $\phi(x_1, x_1)$ durch Umwandlungen geschaffen usw. Die so umgeformten Funktionen, die verschiedene Bezeichnungen haben, sind als ganz voneinander und von der Ursprünglichen verschiedene Funktionen anzusehen.†

2) Einsetzung einer Konstanten a für eine Variable x_k in einer Funktion von n Variablen, wo $n \geq k$. Dadurch wird eine Funktion von $n - 1$ Variablen (bzw., wenn $n = 1$ ist, eine Konstante) geformt. Dies soll in der folgenden Weise bezeichnet werden: für x_k in der ursprünglichen Funktion soll immer a auftreten, für x_i mit $i > k$ soll x_{i-1} auftreten, während die x_i mit $i < k$ unverändert bleiben sollen.

3) Zusammensetzung einer Funktion von n Variablen mit einer von m Variablen, durch Einsetzung von dieser *als Funktion von m Variablen* für x_k ($k \leq n$) in jene. Die so gestaltete Funktion soll in der folgenden Weise bezeichnet werden: Zunächst setzt man in die zweite Funktion y_i für x_i ein, dann setzt man den so umgeformten Ausdruck als ein Ganzes in die erste Funktion in alle die Stellen ein, wo x_k da erscheint, und endlich formt man den resultierenden Ausdruck so um, dass x_i für $i < k$ unverändert bleibt, y_i in x_{k+i-1} übergeht, und x_i für $i > k$ in x_{i+m-1} übergeht. Z. B. durch Einsetzung von $\psi(x_2, x_1, x_1, x_3)$ für x_2 in $\phi(x_3, x_2, x_1, x_2, x_4, x_5, x_6)$ hat man $\phi(x_5, \psi(x_3, x_2, x_2, x_4), x_1, \psi(x_3, x_2, x_2, x_4), x_6, x_7, x_8)$. (Wenn man eine Konstante als eine Funktion von 0 Variablen ansieht, so ist der Fall (2) als Spezialfall im Falle (3) eingeschlossen).

Diese Operationen sind die, mit denen wir uns in diesem Kapitel zu beschäftigen haben. Die Gesamtheit der Ausdrücke, die sie erzeugen, hat die Eigenschaft, dass jede Funktion und jede Aussage, die aus den Grundfunktionen und Grundkonstanten gebildet werden kann, mit einem und nur einem Ausdruck bezeichnet wird. Es kann aber geschehen, dass derselbe Ausdruck in mehreren Weisen durch diese Operationen erzielt wird, also dass ganz verschiedene Operationsprozesse in dem Sinne äquivalent sind, dass sie dasselbe Etwas liefern. Diese Äquivalenzen sind die Eigenschaften, die hier in Betracht kommen.

Diese Ausdrücke lassen sich nun in unsere Schreibweise umformen, wenn wir nur die Variablen behandeln, als ob sie Etwase wären. In der Tat geht ein Ausdruck der Form $f(u_1, u_2, \cdots u_n)$ wegen der Ausdeutung der Anwen-

1) The *transformation* (*Umwandlung*) of one function into a second that differs from the first only in that the variables are numbered differently. This new numbering should satisfy the above condition of consecutiveness, yet it is permitted to assign the same sign to blanks that originally had different signs, but not conversely. E.g. from $\phi(x_1, x_2)$ we can get by transformation $\phi(x_2, x_1)$ and $\phi(x_1, x_1)$, etc. The functions so transformed, that have different notations, are considered entirely different functions from each other and from the original functions.

2) Replacement of a variable x_k by a constant a in a function of n variables, where $n \geq k$. The result of this replacement will be a function of $n - 1$ variables (or, if $n = 1$, a constant.) This will be used in the following way: for x_k in the original function a will be written, for x_i with $i > k$, x_{i-1} will be written, whereas x_i with $i < k$ will be left unchanged.

3) Composition of a function of n variables with one of m variables by the replacement of x_k ($k \leq n$) by *a function of m variables*. The resulting function will be denoted as follows: first one replaces each x_i by y_i in the second function, then one places the resulting expression as a whole in every place where x_k appears, and finally one forms the resulting expression so that x_i for $i < k$ remains unchanged, y_i is replaced by x_{k+i-1} and x_i for $i > k$ is replaced by x_{i+m-1}. E.g., by the replacement of x_2 by $\psi(x_2, x_1, x_1, x_3)$ in

$$\phi(x_3, x_2, x_1, x_2, x_4, x_5, x_6)$$

one has $\phi(x_5, \psi(x_3, x_2, x_2, x_4), x_1, \psi(x_3, x_2, x_2, x_4), x_6, x_7, x_8)$. (If one views a constant as a function of 0 variables, then case (2) is a special case of case (3).)

These operations are those with which we are concerned in this chapter. The collection of expressions that they produce has the property that every function and every statement that can be built from the basic functions and basic constants can be denoted by one and only one expression. But it can happen that the same expression can be obtained in more than one way using these operations, and so different operation-processes are equivalent in the sense that they lead to the same entity. These equivalences are the properties that concern us here.

These expressions may now be transformed into our notation if we treat the variables as if they were entities. In fact, an expression of the form $f(u_1, u_2, \cdots, u_n)$, because of the interpretation of application (as Schönfinkel's function concept),

dung (als Schönfinkelscher Funktionsbegriff) in $(fu_1u_2 \cdots u_n)$ über. Z. B. wird der oben (in 3) betrachtete Ausdruck nach der Umformung

$$(\phi x_5(\psi x_3x_2x_2x_4)x_1(\psi x_3x_2x_2x_4)x_6x_7x_8).$$

Die neuen Ausdrücke sind also Kombinationen von gewissen Grundfunktionen, Grundkonstanten und Variablen. Diese Kombinationen lassen sich aber ferner aus Kombinationen von lauter Variablen erschaffen und zwar dadurch, dass man in eine der letzten für x_1 eine Grundfunktion nach dem obigen Prozess 2) einsetzt, dann für x_2 eine Grundkonstante oder Grundfunktion einsetzt, u. s. w., bis alle die im betreffenden Ausdruek erscheinenden Grundgegenstände eingesetzt sind. Z. B. wird das oben Geschriebene aus

$$(x_1x_7(x_2x_5x_4x_4x_6)x_3(x_2x_5x_4x_4x_6)x_8x_9x_{10})$$

durch Einsetzung von ϕ statt x_1 und ψ statt x_2 erzielt.

Diese letzte Bemerkung läuft darauf hinaus, dass die Kombinationen lauter Variablen Operatoren sind, die aus den gegebenen Funktionen und Konstanten alle möglichen abgeleiteten Funktionen und Konstanten durch Einsetzung, und zwar die Schönfinkelsche,* hervorrufen. In diesem Kapitel zeige ich, dass diese Operatoren nichts wesentlich anderes als eine bestimmte Klasse von Kombinatoren sind. In der Tat setze ich zuerst fest, dass ein Kombinator Y eine Kombination der Variablen $x_1, x_2 \cdots x_n$, nämlich X dann und nur dann *darstellt*, wenn es formal, –d. h. durch behandlung der Variablen als Etwase ohne besondere Eigenschaften–folgt, das

$$\vdash Yx_1x_2 \cdots x_n = X$$

Dann beweise ich die folgenden Hauptsätze:

I. *Wenn ein Kombinator Y eine Kombination von $x_1x_2 \cdots x_n$ darstellt, so stellt er nur eine dar (bewiesen in C1).*

II. *Zu jeder Kombination lauter Variabler gibt es mindestens einen Kombinator Y, der sie darstellt (bewiesen in E1).*

III. *Wenn zwei Kombinatoren Y_1 und Y_2 dieselbe Kombination lauter Variabler darstellen, dann folgt es ohne Gebrauch von Variablen, dass* $\vdash Y_1 = Y_2$ *(bewiesen in E 3)*.†

Aus diesen drei Hauptsätzen folgt leicht, was ich beweisen will. Denn zunächst wird jeder Ausdruck eindeutig in der Form $Yu_1u_2 \cdots u_n$, wo $u_1, u_2 \cdots u_n$ die in dem Ausdruck erscheinenden Grundgegenstände sind,

---

*Vgl. oben unter I A.

† Mit der unwichtigen Beschränkung dass Y_1 und Y_2 eigentlich (s. unten II E) sind.

becomes $(fu_1u_2\cdots u_n)$. E.g. the expression given above (in case 3)) becomes by this transformation

$$(\phi x_5(\psi x_3 x_2 x_2 x_4)x_1(\psi x_3 x_2 x_2 x_4)x_6 x_7 x_8).$$

The new expressions are thus combinations of certain basic functions, basic constants, and variables. But further, these combinations may be reconstructed from combinations of variables only (*Kombinationen von lauter Variablen*):[10] in such a combination one can substitute a basic function for x_1 by the above process 2), then put a basic constant or basic function for x_2, etc., until all the basic objects appearing in the expression concerned have been inserted. E.g., the above expression can be obtained from

$$(x_1 x_7(x_2 x_5 x_4 x_4 x_6)x_3(x_2 x_5 x_4 x_4 x_6)x_8 x_9 x_{10})$$

by putting ϕ for x_1 and ψ for x_2.

It follows from this last remark that the combinations of pure variables are operators, that generate from the given functions and constants all possible derived functions and constants that can be obtained by substitution, including those of Schönfinkel.* In this chapter, I prove that these operators are truly nothing other than a certain class of combinators. In fact, I first say that a combinator Y *represents* a combination of the variables $x_1, x_2, \cdots, x_n$, namely X, if and only if it follows formally,–i.e. through the treatment of the variables as entities without any particular properties,–that
$$\vdash Y x_1 x_2 \cdots x_n = X.$$
Then I prove the following main theorems:

I. *If a combinator Y represents a combination of $x_1, x_2, \cdots, x_n$, then it represents only one such* (proven in C 1).

II. *To each combination of variables only, there is at least one combinator Y which represents it* (proven in E 1).

III. *If two combinators Y_1 and Y_2 represent the same combination of variables only, then it follows without the use of variables that* $\vdash Y_1 = Y_2$ (proven in E 3).†

From these three main theorems what I wish to prove will follow easily. First, every expression will be represented uniquely in the form
$$Y u_1 u_2 \cdots u_n,$$
where $u_1, u_2, \cdots, u_n$ are the basic objects appearing in the expression,

---

*Cf. above under I A.
†With the trivial restriction that Y_1 and Y_2 are pure (*eigentlich*) (cf. II E below).

dargestellt, und weiter lässt sich jeder Substitutionsprozess so durch Kombinatoren definieren, dass aus den Ausdrücken der Form $Yu_1u_2\cdots u_n$ immer neue Ausdrücke derselben Form erzeugt werden.

Dieses Resultat ist freilich nur ein Spezialergebnis der drei Hauptsätze, welche eine gewisse Art von Isomorphismus zwischen Kombinatoren und Kombinationen überhaupt aussagen. Mit dem Kombinator K kommen Kombinationen mit Auslassungen in Betracht. Wir könnten diese ausschliessen, und nur Kombinatoren, die Kombinationen von B, C, W, und I sind, betrachten; aber die allgemeineren Sätze können fast so leicht bewiesen werden als die besonderen, und also habe ich den K behalten. Die Betrachtungen, die den K speziell betreffen, sind unten zuweilen nicht so ausführlich angegeben, wie die anderen.

Eine letzte Bemerkung kommt dazu. In der eben dargestellten Theorie wurden bisher Funktionen von verschiedenen Anzahlen von Variablen unterschieden; dagegen ist jedes Etwas wegen Reg. E eine Funktion beliebig vieler Variablen. Infolgedessen hat man mit jedem Etwas nicht bloss eine bestimmte endliche Anzahl von Leerstellen zu assoziieren, sondern eine unendliche Folge von Leerstellen. Durch irgendeinen der obigen Prozesse wird aber nur eine endliche Anzahl von Leerstellen gestört, also haben diese Folgen einem besonderen Charakter. Dies erklärt die Tatsache, dass von Abschnitt C ab Folgen einer bestimmten Art eine grosse Rolle spielen.

B. DIE GRUNDLEGENDEN DEFINITIONEN UND SÄTZE.

§1. *Die B Sequenz.*
Def. 1. $B_1 \equiv B$; $B_{n+1} \equiv BBB_n$, $(n = 1, 2, 3, \cdots)$.
Def. 2. $B_0 \equiv I$.
SATZ 1. *Wenn X, Y und Z Etwase sind, dann*

$$\vdash B_{n+1}XYZ = B_nX(YZ), \qquad (n = 0, 1, 2, \cdots).$$

Beweis: Für $n = 0$ folgt der Satz aus Def. 1 und 2, Reg. B, und I D Satz 9.
Es sei $n > 0$;

$$\begin{aligned}
\vdash B_{n+1}XYZ &= BBB_nXYZ & \text{(Def. 1, I D Satz 6),}\\
&= B(B_nX)YZ & \text{(Reg. } B\text{),}\\
&= B_nX(YZ), \quad \text{w.z.b.w.} & \text{(Reg. } B\text{).}
\end{aligned}$$

and furthermore every substitution process can be so defined by combinators that out of the expressions of the form $Y u_1 u_2 \cdots u_n$ new expressions of the same form can always be produced.

This result is certainly only a special case of the three main theorems, which generally indicates a certain kind of isomorphism between combinators and combinations. Combinations with omissions must be taken into account with the combinator K. We could exclude these, and consider only combinators that are combinations of B, C, W, and I; but the general theorems can be proved almost as easily as the more restricted ones, and so I keep the combinator K. The considerations that specifically concern K are not given below with as much detail as the others.

A last remark. In the theory as given so far, functions of different numbers of variables have been differentiated; on the other hand, because of Rule E, every entity is a function of arbitrarily many variables. Consequently, one only has to associate with every entity not just a certain finite number of blanks, but an infinite sequence of blanks. By any of the above processes, only a finite number of blanks will be disturbed, so these sequences have a special character. This explains the fact that from section C on, sequences of a special kind play a large role.

B. The Basic Definitions and Theorems

§1. *The B sequence*

Def. 1. $B_1 \equiv B$; $\qquad B_{n+1} \equiv BBB_n$, $\qquad (n = 1, 2, 3, \cdots)$.

Def. 2. $B_0 \equiv I$.

Theorem 1. *If X, Y, and Z are entities, then*
$$\vdash B_{n+1}XYZ = B_n X(YZ), \qquad (n = 0, 1, 2, \cdots).$$
Proof. For $n = 0$ the theorem follows from Def. 1 and 2, Rule B, and I D Theorem 9.

Let $n > 0$;
$$\vdash B_{n+1}XYZ \;=\; BBB_n XYZ \qquad \text{(Def. 1, I D Theorem 6)},$$
$$=\; B(B_n X)YZ \qquad \text{(Rule } B\text{)},$$
$$=\; B_n X(YZ), \quad \text{q.e.d.} \quad \text{(Rule } B\text{)}.$$

SATZ 2. *Wenn* $X, Y, Z_1, Z_2, \cdots, Z_m$ *Etwase sind, dann*

$$\vdash B_{n+m}XYZ_1Z_2\cdots Z_m = B_nX(YZ_1Z_2\cdots Z_m),$$
$$(m = 0, 1, 2, \cdots; n = 0, 1, 2, \cdots).$$

Beweis: Für $m = 0$ klar. Für $m = 1$ folgt der Satz aus Satz 1.

Angenommen, der Satz sei für $m = k$ bewiesen. Dann wird er für $m = k + 1$ bewiesen, wie folgt: aus Satz 1 folgt

$$\vdash B_{n+k+1}XYZ_1\cdots Z_kZ_{k+1} = B_{n+k}X(YZ_1)Z_2Z_3\cdots Z_kZ_{k+1},$$
$$(n = 0, 1, 2, \cdots).$$

Nun in den vorliegenden Satz für $m = k$ setzen wir (YZ) für Y, und Z_{i+1} für Z_i ($i = 1, 2, \cdots, k$). Dann

$$\vdash B_{n+k}X(YZ_1)Z_2Z_3\cdots Z_kZ_{k+1} = B_nX(YZ_1Z_2\cdots Z_kZ_{k+1}),$$
$$(n = 0, 1, 2, \cdots).$$

Aus den letzten zwei Formeln folgt der Satz für $m = k+1$. Durch Wiederholung dieses Prozesses wird der Satz für ein beliebiges m bewiesen.

SATZ 3. *Sind* $X, Y, Z_1, Z_2, \cdots, Z_m$ *Etwase, dann*

$$\vdash B_mXYZ_1Z_2\cdots Z_m = X(YZ_1Z_2\cdots Z_m), \qquad (m = 0, 1, 2, 3, \cdots).$$

Beweis: Man setzt $n = 0$ in Satz 2.

SATZ 4. *Wenn eine Reihe von Etwasen* $X_1, X_2, X_3, \cdots$, *einer Rekursionsformel, näm.,*

$$\vdash X_{n+1} = BX_n \qquad (n = 1, 2, 3, \cdots),$$

erfüllen, dan

$$\vdash X_{n+m} = B_mX_n, \qquad (m = 0, 1, 2, \cdots; n = 1, 2, 3, \cdots).$$

Beweis: Für $m = 0$ oder $m = 1$ klar.
Nun sei der Satz für $m = k$ augenommen. Dann folgt

$$\begin{aligned}
\vdash X_{n+k+1} &= BX_{n+k} & \text{(Hp.)}, \\
&= B(B_kX_n) & \text{(nach diesem Satz für } m = k\text{)}, \\
&= BBB_kX_n & \text{(Reg. } B\text{)}, \\
&= B_{k+1}X_n & \text{(Def. 1)}.
\end{aligned}$$

Also kann der Satz für ein beliebiges m bewiesen werden.

SATZ 5. *Ist* X *ein Etwas, dann*

$$\vdash B_m(B_nX) = B_{m+n}X \qquad (m, n = 0, 1, 2, 3, \cdots).$$

Theorem 2. If $X, Y, Z_1, Z_2, \cdots, Z_m$ are entities, then
$$\vdash B_{n+m}XYZ_1Z_2\cdots Z_m = B_nX(YZ_1Z_2\cdots Z_m),$$
$$(m = 0, 1, 2, \cdots; n = 0, 1, 2, \cdots).$$

Proof. For $m = 0$ this is clear. For $m = 1$ this holds by Theorem 1.

Assume that the theorem has been proved for $m = k$. Then it will be proved for $m = k+1$ as follows: from Theorem 1 follows
$$\vdash B_{n+k+1}XYZ_1Z_2\cdots Z_kZ_{k+1} = B_{n+k}X(YZ_1)Z_2Z_3\cdots Z_kZ_{k+1},$$
$$(n = 0, 1, 2, \cdots).$$

Now in the theorem for $m = k$, we put (YZ_1) for Y and Z_{i+1} for Z_i ($i = 1, 2, \cdots, k$). Then
$$B_{n+k}X(YZ_1)Z_2Z_3\cdots Z_kZ_{k+1} = B_nX(YZ_1Z_2\cdots Z_kZ_{k+1}),$$
$$(n = 0, 1, 2, \cdots).$$

From the last two formulas the theorem follows for $m = k+1$. By repeting these processes the theorem will be proved for any m.

Theorem 3. If $X, Y, Z_1, Z_2, \cdots, Z_m$ are entities, then
$$\vdash B_mXYZ_1Z_2\cdots Z_m = X(YZ_1Z_2\cdots Z_m), \qquad (m = 0, 1, 2, 3, \cdots).$$

Proof. Let $n = 0$ in Theorem 2.

Theorem 4. If a sequence of entities $X_1, X_2, X_3, \cdots$ satisfies a recursion formula, namely
$$\vdash X_{n+1} = BX_n \qquad (n = 1, 2, 3, \cdots),$$
then
$$\vdash X_{n+m} = B_mX_n, \qquad (m = 0, 1, 2, \cdots; n = 1, 2, 3, \cdots).$$

Proof. For $m = 0$ or $m = 1$ this is clear.

Now let the theorem be assumed for $m = k$. Then
$$\vdash X_{n+k+1} = BX_{n+k} \quad \text{(Hp.)},$$
$$= B(B_kX_n) \quad \text{(by this theorem for } m = k\text{)},$$
$$= BBB_kX_n \quad \text{(rule } B\text{)},$$
$$= B_{k+1}X_n \quad \text{(Def. 1)}.$$

So the theorem can be proved for any m.

Theorem 5. If X is an entity, then
$$\vdash B_m(B_nX) = B_{m+n}X, \qquad (m, n = 0, 1, 2, 3, \cdots).$$

Beweis: Für $n = 0$ klar, weil $\vdash B_0 X = X$. Sonst setzt man:

$$X_n \equiv B_n X, \qquad (n = 1, 2, 3, \cdots),$$

dann

$$\begin{aligned} \vdash X_{n+1} &= B_{n+1} X = B B B_n X &\text{(Def. 1)}, \\ &= B(B_n X) = B X_n &\text{(Reg. } B\text{)}. \end{aligned}$$

Der Satz folgt dann aus Satz 4.

§2. Die Identitätsfunktion.

SATZ 1. $\vdash B_m I = I$, $\qquad (m = 0, 1, 2, \cdots)$.

Beweis: In §1 Satz 4 setzen wir $X_n \equiv I$. Dann

$$\vdash X_{n+1} = B X_n \qquad \text{(Ax. } I_2\text{)}.$$

Daher folgt der Satz aus §1 Satz 4.

SATZ 2. *Wenn X ein Etwas ist, dann*

$$\vdash B_m I X = X, \qquad (m = 1, 2, 3, \cdots).$$

Beweis : $\begin{aligned} \vdash B_m I X &= I X &\text{(Satz 1)}, \\ &= X, &\text{(I D Satz 9)}. \end{aligned}$

SATZ 3. $\vdash CBI = I$.

Beweis : $\begin{aligned} \vdash CBI &= B(BI)I &\text{(Ax. } I_1\text{)} \\ &= BII = II &\text{(Ax. } I_2\text{)} \\ &= I. &\text{(I D Satz 9)}. \end{aligned}$

SATZ 4. *Wenn X ein Etwas ist, dann gilt* $\vdash BXI = X$.

Beweis : $\begin{aligned} \vdash BXI &= CBIX &\text{(Reg. } C\text{)}, \\ &= IX &\text{(Satz 3)}, \\ &= X &\text{(I D Satz 9)}. \end{aligned}$

SATZ 5. *Def. 1 von §1 gilt auch, wenn $n = 0$, d. h.*

$$\vdash B_1 = B B B_0.$$

Beweis: Klar aus Satz 4 und §1, Def. 2.

§3. Die C, W und K Sequenzen.

Def. 1. $C_1 \equiv C$; $C_{n+1} \equiv B C_n$, $\qquad (n = 1, 2, 3, \cdots)$.

SATZ 1. $\vdash B_m C_n = C_{m+n}$ $\qquad (m = 0, 1, 2, \cdots; n = 1, 2, 3, \cdots)$.

Proof. For $n=0$ this is clear, because $\vdash B_0 X = X$. Otherwise let

$$X_n \equiv B_n X, \qquad (n=1,2,3,\cdots),$$

and then

$$\begin{aligned}\vdash X_{n+1} &= B_{n+1}X = BBB_n X &\text{(Def. 1)},\\&= B(B_n X) = BX_n &\text{(Rule } B\text{)}.\end{aligned}$$

The theorem then follows from Theorem 4.

§2. The identity function.

Theorem 1. $\vdash B_m I = I$, $\qquad (m=0,1,2,\cdots)$.

Proof. In §1 Theorem 4, let $X_n \equiv I$. Then

$$\vdash X_{n+1} = BX_n \qquad \text{(Ax. } I_2\text{)}.$$

From this the theorem follows by §1 Theorem 4.

Theorem 2. If X is an entity, then

$$\vdash B_m I X = X, \qquad (m=1,2,3,\cdots).$$

Proof.
$$\begin{aligned}\vdash B_m I X &= IX &\text{(Theorem 1)},\\&= X, &\text{(I D Theorem 9)}.\end{aligned}$$

Theorem 3. $\vdash CBI = I$.

Proof.
$$\begin{aligned}\vdash CBI &= B(BI)I &\text{(Ax. } I_1\text{)}\\&= BII = II &\text{(Ax. } I_2\text{)}\\&= I. &\text{(I D Theorem 9)}.\end{aligned}$$

Theorem 4. If X is an entity, then $\vdash BXI = X$.

Proof.
$$\begin{aligned}\vdash BXI &= CBIX &\text{(Rule } C\text{)},\\&= IX &\text{(Theorem 3)},\\&= X &\text{(I D Theorem 9)}.\end{aligned}$$

Theorem 5. *Def. 1 of §1 holds also when* $n=0$, *i.e.*

$$\vdash B_1 = BBB_0.$$

Proof. Clear by Theorem 4 and §1 Def. 2.

§3. The C, W, and K sequences.

Def. 1. $C_1 \equiv C$; $C_{n+1} \equiv BC_n$, $\qquad (n=1,2,3,\cdots)$.

Theorem 1. $\vdash B_m C_n = C_{m+n}$, $\qquad (m=0,1,2,\cdots; n=1,2,3,\cdots)$.

Beweis: Folgt unmittelbar aus Def. 1 und §1 Satz 4.

SATZ 2. *Wenn* $X_0, X_1, X_2, \cdots, X_n, X_{n+1}$, *Etwase sind, dann*

$$\vdash C_n X_0 X_1 X_2 \cdots X_n X_{n+1} = X_0 X_1 X_2 \cdots X_{n-1} X_{n+1} X_n.$$

Beweis : $\vdash C_n X_0 X_1 \cdots X_{n-1} X_n X_{n+1}$
$= B_{n-1} C_1 X_0 X_1 \cdots X_{n-1} X_n X_{n+1}$ (Satz 1),
$= C_1(X_0 X_1 X_2 \cdots X_{n-1}) X_n X_{n+1}$ (§1 Satz 3),
$= X_0 X_1 X_2 \cdots X_{n-1} X_{n+1} X_n,$ (Reg. C).

Def. 2. $W_1 \equiv W;$ $\quad W_{n+1} \equiv BW_n \quad\quad (n = 1, 2, 3, \cdots)$.

SATZ 3. $\vdash B_m W_n = W_{m+n},\quad\quad (m = 0, 1, 2, \cdots; n = 1, 2, 3, \cdots).$

Beweis: Folgt unmittelbar aus Def. 2 und §1, Satz 4.

SATZ 4. *Wenn* $X_0, X_1, X_2, \cdots X_n$ *Etwase sind, so gilt*

$$\vdash W_n X_0 X_1 \cdots X_n = X_0 X_1 \cdots X_{n-1} X_n X_n.$$

Beweis: Wie der von Satz 2, mit Gebrauch von Reg. W anstatt Reg. C.

Def. 3. $K_1 \equiv K,\ K_{n+1} \equiv BK_n,\quad\quad (n = 1, 2, 3, \cdots).$

SATZ 5. $\vdash B_m K_n = K_{m+n},\quad (m = 1, 2, 3, \cdots; n = 1, 2, 3, \cdots).$

SATZ 6. *Wenn* $X_0, X_1, X_2, \cdots, X_n$ *Etwase sind, so gilt*

$$K_n X_0 X_1 X_2 \ldots X_n = X_0 X_1 X_2 \cdots X_{n-1}.$$

Beweis: Wie der von Satz 2 mit Gebrauch von Reg. K anstatt Reg. C.

§4. *Das zusammengesetzte Produkt.*

Def. 1. Wenn X und Y Etwase sind, $(X \cdot Y) \equiv BXY$.

SATZ 1. *Sind X, Y, Z Etwase, dann* $\vdash (X \cdot Y)Z = X(YZ).$

Beweis: Folgt aus Def. 1 und Reg. B.

Def. 2. Wenn X und Y Bezeichnungen für Etwase sind, und keine von den beiden von der Form $(U = V)$ oder $(U \cdot V)$ ist, so dürfen wir in einer Bezeichnung wie $X \cdot Y$ die äusseren Klammern von X und Y fortlassen.

SATZ 2. *Sind X und Y Etwase, dann gilt*

$$\vdash B(X \cdot Y) = BX \cdot BY.$$

Proof. This follows immediately by Def. 1 and §1 theorem 4.

Theorem 2. *If $X_0, X_1, X_2, \cdots, X_n, X_{n+1}$ are entities, then*
$$\vdash C_n X_0 X_1 X_2 \cdots X_n X_{n+1} = X_0 X_1 X_2 \cdots X_{n-1} X_{n+1} X_n.$$

Proof. $\vdash C_n X_0 X_1 \cdots X_{n-1} X_n X_{n+1}$
$\quad = B_{n-1} C_1 X_0 X_1 \cdots X_{n-1} X_n X_{n+1}$ (Theorem 1),
$\quad = C_1 (X_0 X_1 \cdots X_{n-1}) X_n X_{n+1}$ (§1 Theorem 3),
$\quad = X_0 X_1 X_2 \cdots X_{n-1} X_{n+1} X_n$ (Rule C).

Def. 2. $W_1 \equiv W$; $W_{n+1} \equiv BW_n$, $\quad (n = 1, 2, 3, \cdots)$.

Theorem 3. $\vdash B_m W_n = W_{m+n}$, $\quad (m = 0, 1, 2, \cdots; n = 1, 2, 3, \cdots)$.

Proof. This follows immediately by Def. 2 and §1 Theorem 4.

Theorem 4. *If $X_0, X_1, X_2, \cdots, X_n$ are entities, then*
$$\vdash W_n X_0 X_1 \cdots X_n = X_0 X_1 \cdots X_{n-1} X_n X_n.$$

Proof. Like that of Theorem 2, use Rule W instead of Rule C.

Def. 3. $K_1 \equiv K$, $K_{n+1} \equiv BK_n$, $\quad (n = 1, 2, 3, \cdots)$.

Theorem 5. $\vdash B_m K_n = K_{m+n}$, $\quad (m = 1, 2, 3, \cdots; n = 1, 2, 3, \cdots)$.

Theorem 6. *If $X_0, X_1, X_2, \cdots, X_n$ are entities, then*
$$K_n X_0 X_1 X_2 \ldots X_n = X_0 X_1 X_2 \cdots X_{n-1}.$$

Proof. Like that of Theorem 2, use Rule K instead of Rule C.

§4. *The composite product.*

Def. 1. If X and Y are entities, $(X \cdot Y) \equiv BXY$.

Theorem 1. *If X, Y, and Z are entities, then*
$$\vdash (X \cdot Y) Z = X(YZ).$$

Proof. This follows from Def. 1 and Rule B.

Def. 2. If X and Y denote entities, and neither of them has the form $(U = V)$ or $(U \cdot V)$, then we may in a notation such as $X \cdot Y$ omit the outermost parentheses of X and Y.

Theorem 2. *If X and Y are entities, then*
$$\vdash B(X \cdot Y) = BX \cdot BY.$$

Beweis:

$$\begin{align}
(1)\quad \vdash B(X \cdot Y) &= B(BXY) && \text{(Def. 1)}, \\
&= B_2BBXY && \text{(§1, Satz 3)}. \\
(2)\quad \vdash BX \cdot BY &= B(BX)(BY) && \text{(Def. 1 und 2)}, \\
&= B_2X(BY) && \text{(§1, Satz 5, } m=n=1), \\
&= B_3XBY && \text{(§1, Satz 2, } m=1, n=2), \\
&= CB_3BXY && \text{(Reg. } C.)
\end{align}$$

Nun aus §1, Def. 1,

$$\vdash CB_3B = C(BB(BBB))B,$$

daher aus Ax. B

$$\begin{align}
(3)\quad \vdash CB_3BXY &= B(BB)BXY \\
&= B_2BBXY && \text{(§1, Satz 5)}.
\end{align}$$

Aus (1), (2), and (3), wird der satz bewiesen.

SATZ 3. *Das Proclukt $(X \cdot Y)$ ist assoziativ, d.h., wenn X, Y, Z Etwase sind, so gilt*

$$\vdash X \cdot (Y \cdot Z) = (X \cdot Y) \cdot Z.$$

Beweis:

$$\begin{align}
\vdash X \cdot (Y \cdot Z) &= B(X \cdot Y)Z && \text{(Def. 1)}, \\
&= (BX \cdot BY)Z && \text{(Satz 2)}, \\
&= BX(BYZ) && \text{(Satz 1)}, \\
&= X \cdot (Y \cdot Z) && \text{(Def. 1)}.
\end{align}$$

SATZ 4. *Wenn X ein Etwas ist, so gelten*

1) $\vdash I \cdot X = X$,

2) $\vdash X \cdot I = X$.

Beweis: Folgt aus Def. 1 und §2, Sätze 2 und 4.

SATZ 5. $\vdash B_m \cdot B_n = B_{m+n}$, $\qquad (m, n = 1, 2, 3, \cdots)$.

Beweis: Für $m = 1$ folgt aus Def. 1 und §1, Def. 1. Ist nun der Satz für $m = k$ bewiesen, dann folgt

$$\begin{align}
\vdash B_{k+1} \cdot B_n &= (B \cdot B_k) \cdot B_n && \text{(nach diesem Satze für } m=1), \\
&= B \cdot (B_k \cdot B_n) && \text{(Satz 3)}, \\
&= B \cdot B_{n+k} && \text{(nach diesem Satze für } m=k), \\
&= B_{n+k+1} && \text{(nach diesem Satze für } m=1).
\end{align}$$

Dadurch wird der Satz für ein beliebiges m bewiesen.

Proof.

(1) $\vdash B(X \cdot Y) = B(BXY)$ (Def. 1),
$= B_2BBXY$ (§1, Theorem 3).

(2) $\vdash BX \cdot BY = B(BX)(BY)$ (Def. 1 and 2),
$= B_2X(BY)$ (§1, Theorem 5, $m = n = 1$),
$= B_3XBY$ (§1, Theorem 2, $m = 1, n = 2$),
$= CB_3BXY$ (Rule C.)

Now by §1, Def. 1,

$$\vdash CB_3B = C(BB(BBB))B,$$

from which by Ax. B

(3) $\vdash CB_3BXY = B(BB)BXY$
$= B_2BBXY$ (§1, Theorem 5).

By (1), (2), and (3), the theorem is proved.

Theorem 3. *The product $(X \cdot Y)$ is associative, i.e., if X, Y, and Z are entities, then*

$$\vdash X \cdot (Y \cdot Z) = (X \cdot Y) \cdot Z.$$

Proof. $\vdash (X \cdot Y) \cdot Z = B(X \cdot Y)Z$ (Def. 1),
$= (BX \cdot BY)Z$ (Theorem 2),
$= BX(BYZ)$ (Theorem 1),
$= X \cdot (Y \cdot Z)$ (Def. 1).

Theorem 4. *If X and Y are entities, then*

1) $\vdash I \cdot X = X$,

2) $\vdash X \cdot I = X$.

Proof. This follows from Def. 1 and §2, Theorems 2 and 4.

Theorem 5. $\vdash B_m \cdot B_n = B_{m+n}$, $\qquad (m, n = 1, 2, 3, \cdots)$.

Proof. For $m = 1$ this follows by Def. 1 and §1, Def. 1. If the theorem has been proved for $m = k$, then follows

$\vdash B_{k+1} \cdot B_n = (B \cdot B_k) \cdot B_n$ (by this theorem for $m = 1$),
$= B \cdot (B_k \cdot B_n)$ (Theorem 3),
$= B \cdot B_{n+k}$ (by this theorem for $m = k$),
$= B_{n+k+1}$ (by this theorem for $m = 1$).

So the theorem is proved for arbitrary m.

SATZ 6. *Sind X und Y Etwase, dann gilt,*

$$\vdash B_m(X \cdot Y) = B_m X \cdot B_m Y, \qquad (m = 1, 2, 3, \cdots).$$

Beweis: Für $m = 1$ ist der Satz mit Satz 2 identisch. Nun sei der Satz für $m = k$ angenommen, dann wird er für $m = k+1$ bewiesen, wie folgt:

$$\begin{aligned}
\vdash B_{k+1}(X \cdot Y) &= B(B_k(X \cdot Y)) && (\S 1, \text{Satz } 5), \\
&= B(B_k X \cdot B_k Y) && (\text{nach diesem Satze für } m = k), \\
&= B(B_k X) \cdot B(B_k Y) && (\text{Satz } 2), \\
&= B_{k+1} X \cdot B_{k+1} Y, && (\S 1, \text{Satz } 5).
\end{aligned}$$

SATZ 7. $\vdash B_m B_n \cdot B_m B_p = B_m B_{n+p}, \qquad (m, n, n = 0, 1, 2, \cdots).$

Beweis: Folgt aus Sätzen 4,* 5, 6.

Def. 3. Sind $X_1, X_2, \cdots, X_n$ Etwase, dann

$$(X_1 \cdot X_2 \cdot X_3 \cdot \cdots \cdot X_n) \equiv (\cdots(((X_1 \cdot X_2) \cdot X_3) \cdot X_4) \cdots X_n).$$

Festsetzung: Da ich die Assoziativität des Produkts $(X \cdot Y)$ bewiesen habe, brauche ich diese Tatsache nicht immer explizit hervortreten zu lassen. Im Gegenteil werde ich die Form der Def. 3 benutzen, und dann habe ich ein Recht, Klammern nach Belieben einzusetzen. Die Def. 3 setzt aber an sich die Assoziativität nicht voraus, und wird benutzt, wo die Assoziativität nicht behauptet ist (z. B. s. die nächste Nummer).

§5. *Die Axiome in der neuen Darstellungsweise.*
SATZ. *Wenn man die Definitionen von §§1-4 berücksichtigt, nehmen die kombinatorischen Axiome die folgende Gestalt an:*

a. *Kommutative Axiome.*

Ax. B. $\quad \vdash CB_3 B = BB \cdot B.$
Ax. C. $\quad \vdash CB_3 C = BC \cdot B_2.$
Ax. W. $\quad \vdash CB_2 W = BW \cdot B_2.$
Ax. K. $\quad \vdash CB_2 K = BK \cdot I.$
Ax. I_1. $\quad \vdash CBI = BI \cdot I.$

b. *Transmutative axiome.*

Ax. (BC). $\quad \vdash B_1 \cdot C_1 = C_2 \cdot C_1 \cdot BB.$
Ax. (BW). $\quad \vdash B_1 \cdot W_1 = W_2 \cdot W_1 \cdot C_2 \cdot B_2 B \cdot B.$
Ax. (BK). $\quad \vdash B_1 \cdot K_1 = K_1 \cdot K_1.$

---

* Satz 4 ist nur wenn $n = 0$ oder $p = 0$ notwendig.

Theorem 6. *If X and Y are entities, then*

$$\vdash B_m(X \cdot Y) = B_m X \cdot B_m Y, \qquad (m = 1, 2, 3, \cdots).$$

Proof. For $m = 1$ the theorem is identical with Theorem 2. Now let the theorem for $m = k$ be assumed; then it will be proved for $m = k+1$ as follows:

$$\begin{aligned}
\vdash B_{k+1}(X \cdot Y) &= B(B_k(X \cdot Y)) && (\S 1, \text{Theorem 5}), \\
&= B(B_k X \cdot B_k Y) && (\text{by this theorem for } m{=}k), \\
&= B(B_k X) \cdot B(B_k Y) && (\text{Theorem 2}), \\
&= B_{k+1} X \cdot B_{k+1} Y, && (\S 1, \text{Theorem 5}).
\end{aligned}$$

Theorem 7. $\vdash B_m B_n \cdot B_m B_p = B_m B_{n+p}, \qquad (m, n, p = 0, 1, 2, \cdots)$

Proof. This follows by Theorems 4*, 5, and 6.

Def. 3. If $X_1, X_2, \cdots, X_n$ are entities, then

$$(X_1 \cdot X_2 \cdot X_3 \cdot \cdots \cdot X_n) \equiv (\cdots (((X_1 \cdot X_2) \cdot X_3) \cdot X_4) \cdot \cdots \cdot X_n).$$

Convention. Since I have proved the associativity of the product $(X \cdot Y)$, I may use this fact without explicit mention. Quite the reverse, I will use the form of Def. 3, and then I have the right to insert parentheses as I please. But Def. 3 does not assume associativity, and it will be used where associativity does not hold (e.g., see the next section).

§5. *The axioms in a new manner of presentation.*

Theorem. *When the definitions of §§1–4 are taken into account, the combinatory axioms take the following form:*

a. *Commutative axioms.*

$$\begin{aligned}
\text{Ax. } B. \quad &\vdash CB_3 B = BB \cdot B. \\
\text{Ax. } C. \quad &\vdash CB_3 C = BC \cdot B_2. \\
\text{Ax. } W. \quad &\vdash CB_2 W = BW \cdot B_2. \\
\text{Ax. } K. \quad &\vdash CB_2 K = BK \cdot I. \\
\text{Ax. } I_1. \quad &\vdash CBI = BI \cdot I.
\end{aligned}$$

b. *Transmutative axioms.*

$$\begin{aligned}
\text{Ax. } (BC). \quad &\vdash B_1 \cdot C_1 = C_2 \cdot C_1 \cdot BB. \\
\text{Ax. } (BW). \quad &\vdash B_1 \cdot W_1 = W_2 \cdot W_1 \cdot C_2 \cdot B_2 B \cdot B. \\
\text{Ax. } (BK). \quad &\vdash B_1 \cdot K_1 = K_1 \cdot K_1.
\end{aligned}$$

---

*Theorem 4 is only needed when $n = 0$ or $p = 0$.

Ax. $(CC)_1$. $\vdash C_1 \cdot C_1 = B_2 I$.
Ax. $(CC)_2$. $\vdash C_2 \cdot C_1 \cdot C_2 = C_1 \cdot C_2 \cdot C_1$.
Ax. (CW). $\vdash C_1 \cdot W_1 = W_2 \cdot C_1 \cdot C_2$.
Ax. (CK). $\vdash C_1 \cdot K_1 = K_2$.
Ax. (WC). $\vdash W_1 \cdot C_1 = W_1$.
Ax. (WW). $\vdash W_1 \cdot W_1 = W_1 \cdot W_2$.
Ax. (WK). $\vdash W_1 \cdot K_1 = BI$.

c. *Andere axiome.*

Ax. I_2. $\vdash BI = I$.

Bemerkungen über die Bedeutung dieser Axiomen. Diese Axiomen haben einen Vertauschbarkeitscharakter, d. h., sie sagen aus, dass gewisse Ausdrücke in zwei Weisen gebildet werden können. Z. B. kann der Ausdruck

$$x_0 x_3 (x_1 x_2)$$

aus dem Ausdruck

$$x_0 x_1 x_2 x_3$$

in den folgenden zwei Weisen gebildet werden: 1), man setzt zunächst die Klammern ein, und fürt dann eine Vertauschung aus, oder 2), man macht zuerst eine Vertauschung und dann eine Einsetzung von Klammern. Diese zwei Konstructionsweisen entsprechen den zwei Darstellungen

$$(B \cdot C) x_0 x_1 x_2 x_3$$
$$(C_2 \cdot C_1 \cdot BB) x_0 x_1 x_2 x_3$$

die nach Ax. (BC) gleich sind. Ebenso sagen die sämtlichen tramsmutativen Axiomen eine Vertauschbarkeit aus. Dass die kommutativen Axiomen in dieser Hinsicht nicht wesentlich verschieden sind, folgt aus II D2, Satz 1 (s. umten).

Der Unterschied zwischen den kommutativen und transmutativen Axiomen liegt darin, dass die Kombinatoren, deren Vertauschbarkeit durch jene Axiomen ausgesagt wird, nicht miteinander übergreifen. Um dies zu erklären, setzen wir voraus, dass für einen Kombinator X gilt

$$\vdash X x_0 x_1 x_2 \cdots x_m = x_0 y_1 y_2 \cdots y_n^*$$

wo $y_1, y_2, \cdots, y_n$ Kombinationen von $x_1, x_2, \cdots, x_m$ sind. Ist Y irgendein

---

* Diese Gleichung ist in dem Sinne zu verstehen, dass die Variablen wie beliebige Etwase betrachtet werden.

Ax. $(CC)_1$. ⊢ $C_1 \cdot C_1 = B_2 I$.
Ax. $(CC)_2$. ⊢ $C_2 \cdot C_1 \cdot C_2 = C_1 \cdot C_2 \cdot C_1$.
Ax. (CW). ⊢ $C_1 \cdot W_1 = W_2 \cdot C_1 \cdot C_2$.
Ax. (CK). ⊢ $C_1 \cdot K_1 = K_2$.
Ax. (WC). ⊢ $W_1 \cdot C_1 = W_1$.
Ax. (WW). ⊢ $W_1 \cdot W_1 = W_1 \cdot W_2$.
Ax. (WK). ⊢ $W_1 \cdot K_1 = BI$.

c. *Other axioms.*

Ax. I_2. ⊢ $BI = I$.

Remarks on the meaning of these axioms. These axioms have a transpositional character (*Vertauschenbarkeitscharakter*), i.e., they say that certain expressions can be constructed in two ways. E.g., the expression

$$x_0 x_3 (x_1 x_2)$$

can be constructed from the expression

$$x_0 x_1 x_2 x_3$$

in the following ways: 1) one inserts the parentheses and then transposes, or 2) one transposes and then inserts the parentheses. These two methods of construction correspond to the two expressions

$$(B \cdot C) x_0 x_1 x_2 x_3$$
$$(C_2 \cdot C_1 \cdot BB) x_0 x_1 x_2 x_3,$$

which are equal by Ax. (BC). Thus each and every one of the transmutative axioms indicates a transposition. That the commutative axioms are not really different in this view follows from II D 2 Theorem 1 (see below).

The difference between the commutative and transmutative axioms lies in this, that the combinators whose transpositions are asserted by the former axioms do not overlap with each other. To explain this, we assume that for a combinator X,

$$⊢ X x_0 x_1 x_2 \cdots x_m = x_0 y_1 y_2 \cdots y_n †$$

[11]where $y_1, y_2, \cdots, y_n$ are combinations of $x_1, x_2, \cdots, x_m$. If Y is any combinator

---

†This equation is to be understood in the sense that the variables can denote any given entities.

Kombinator der Form $B_m Z$, so ist es ersichtlich, dass Y auf Variablen wirkt, die durch X nicht gestört werden, und umgekehrt; also muss eine Gleichung der Form

$$B_m Z \cdot X = X \cdot B_n Z$$

inhaltlich richtig sein. Diese Gleichung sagt aber eine Vertauschbarkeit derjenigen Art aus, welche aus den kommutativem Axiomen folgt. Die transmutativen Axiomen haben aber den entgegengesetzten Character. Z. B. wirken die zwei Kombinatoren B und C, die auf der linken Seite des Ax. (BC) erscheinen, auf dieselben Variablen, näm. x_1 und x_2.

Das Ax I_2 ist ein Axiom ganz anderer Art. Seine wesentliche Bedeutung ist dass es uns ermöglicht, die Gleichheit von Kombinatoren zu beweisen, die im Sinne von II C 1 (unten) derselben Folge von lauter Variablen, aber mit verschiedenen Ordnungen entsprechen—was sonst unmöglich wäre (II C 1, Satz 6). D. h., dieses Axiom sagt aus, dass ein Kombinator unabhängig davon zu verstehen ist, wie viele Variablen hinzuzufügen sind, damit eine Reduktion auf eine Kombination von lauter Variablen sich vollzieht.

of the form $B_m Z$, it is evident that Y works on variables which are not disturbed by X, and conversely; thus an equation of the form

$$B_m Z \cdot X = X \cdot B_n Z$$

must be contensively correct (*inhaltliche richtig*).[12] These equations state a transposition of the kind which follows from the commutative axioms. But the transmutative axioms have the opposite character. E.g., the combinators B and C, which appear on the left side of Ax. (BC), work on the same variables, namely x_1 and x_2.

Ax. I_2 is an axiom of an entirely different kind. Its essential meaning is that it makes it possible for us to prove the equality of combinators which in the sense of II C 1 (below) correspond to the same row of pure variables but with different orders–which would otherwise be impossible (II C 1 Theorem 6). I.e., this axiom asserts that a combinator can be understood independently of how many variables are added to obtain a reduction to a combination of variables only.

Grundlagen der kombinatonschen Logik.

TEIL II.*

von H. B. Curry.

C. Darstellung Der Kombinationen Durch Kombinatoren In Der Normalform.

In diesem Abschnitte gebrauchen wir gewisse Zeichen, die wir Variablen nennen wollen. Diese Variablen sind nur ein Hilfsmittel, womit wir zeigen können, dass eine gewisse Art von Vollständigkeit und Verträglichkeit des Grundgerüstes vorliegt. Sie sind nicht als Ableitungen des Grundgerüstes anzusehen. Die Ausführungen dieses Abschnitts haben daher mit der formalen Darstellung nichts zu tun, sondern sie betreffen die Verwandtschaft zwischen dieser und der gewöhnlichen Logik. Diese Variablen sind als Etwase ohne besondere Eigenschaften zu behandeln.

Die Hauptergebnisse dieses Abschnitts sind die letzten Sätze von §1 und §5. Unter den ersten kommt der Hauptsatz I von Abschnitt A vor; dagegen macht §5, Satz 2 den Kern des Hauptsatzes II aus.

§1. *Allgemeines über Reduktion und Entsprechen; ihre Eindeutigkeit.*

Festsetzung 1. In dem Folgenden betrachten wir Ausdrücke, die aus gewissen Variablen $x_1, x_2, x_3, \cdots$ und Etwase formal aufgebaut werden, d. h. so dass die Variablen als Etwase ohne besondere Eigenschaften behandelt werden. Auf solche Ausdrücke werden die vorhergehenden Festsetzungen und Definitionen ausgedehnt.

Festsetzung 2. Wir betrachten nun ein X, das eine Kombination von Kombinatoren und Variablen ist. Wir nehmen an, dass X mit den nach I C, Def. 1, erlaubten Auslassungen von Klammern geschrieben ist, und dass alle die anderen in den vorigen Abschnitten definierten Bezeichnungen durch ihre Definitionen ersetzt sind. Dann ist X von der Form $(X_0 X_1 X_2 \cdots X_n)$, wo X_0 entweder B, C, W, K oder eine Variable x_k ist, und die X_i für $i > 0$ Ausdrücke von derselben Form wie X sind.

Inbezug auf einen solchen X setzen wir zwei Arten von Reduktionsprozessen fest, wie folgt:

1.) Wenn X_0, B, C, W, oder K ist und n gross genug ist, so dürfen wir für

---

*Teil I erschien in diesem Journal, Bd. 52 (1930), S. 509-536.

C. Representation of Combinations via Combinators in Normal Form.[13]

In this section we use certain symbols which we will call variables.[14] These variables are only an expedient with which we may show a kind of completeness and consistency of the foundations presented above. They are not to be seen as derivatives (*Ableitungen*) of the primitive frame. The results of this section therefore have nothing to do with the formal presentation, but they deal with the affinity between this and ordinary logic. These variables are to be taken as entities without any special properties.

The chief results of this section are the last theorems of §1 and §5. Under the first heading is Theorem I of section A; on the other hand, §5 Theorem 2 forms the core of Theorem II.

§1. *Generalities on reduction and equivalence; their explicitness*

Convention 1. In what follows, we use expressions that may be formally constructed from certain variables $x_1, x_2, x_3, \cdots$ and entities, i.e., so that the variables can be taken as entities without any special properties. For such expressions, all the previous conventions and definitions continue to hold.

Convention 2. We now consider an X which is a combination of combinators and variables. We assume that X is written so that parentheses are omitted as permitted by I C Def. 1, and that all the other symbols defined in earlier sections are replaced by their definitions. Then X is of the form $(X_0 X_1 X_2 \cdots X_n)$, where X_0 is either B, C, W, K, or a variable x_k and X_i for $i > 0$ are expressions of the same form as X.

With regard to such an X, we fix two kinds of reduction processes as follows:[15]

1.) If X_0 is B, C, W, or K and if n is large enough, we may replace $X_0 X_1 X_2$

$X_0X_1X_2$ bzw. $X_0X_1X_2X_3$ sein Äquivalent nach der betreffenden Regel B, C, W oder K ersetzen, z. B., wenn $X_0 \equiv B$,* so haben wir $X_1(X_2X_3)X_4\cdots X_n$ anstatt $X_0X_1X_2X_3X_4\cdots X_n$. Eine solche Ersetzung soll ein *Reduktionsprozess erster Art* heissen.

2.) Es mag sein, dass ein Bestandteil von X (d. h. ein eingeklammerter in X erscheinender Ausdruck) durch einen Reduktionsprozess erster Art umgeformt werden kann. Eine solche Umformung soll ein *Reduktionsprozess zweiter Art* heissen, wenn ein Reduktionsprozess erster Art sowohl für den Gesamtausdruck wie auch für jeden Teilausdruck, der den betrefenden einschliesst oder links von ihm steht, unmöglich ist.

Festsetzung 3. Ein Ausdruck X *reduziert* sich auf einen anderen Y, wenn durch Anwendung dieser Prozesse X in Y umgeformt wird, und zwar im *ersten Sinne*, wenn nur Prozesse erster Art nötig sind, und im *zweiten Sinne*, wenn auch Prozesse zweiter Art nötig sind. Dass X sich auf Y reduziert wird auch durch das Zeichen $X \overset{\sim}{=} Y$ ausgedrückt.

Festsetzung 4. Es sei ein Ausdruck X_m gegeben, der x_m, aber keine $x_n, n > m$, enthält, und der ferner nicht von der Form $X_{m-1}x_m$ ist. Dann denken wir an die unendliche Zeichenfolge, welche entsteht, wenn man rechts von X_m die Variablen $x_{m+1}, x_{m+2}, \cdots$ ad infin. setzt; diese heisst die durch X_m bestimmte *Folge*. Der Teil dieser Folge, welcher einem bestimmten $x_n, n > m$, vorangeht, ist ein Ausdruck, der ein *Abschnitt* der Folge heisst. Also ist X_m selbst ein Abschnitt der durch ihn bestimmten Folge.

Festsetzung 5. Ein Ausdruck X enthält eine Variable x_m *wesentlich*, wenn für $n >$ den Index irgendeiner in X erscheinenden Variablen, und für $p \geqq 0$, in der Reduktion von $(Xx_nx_{n+1}\cdots x_{n+p})$ die Variable x_m nie ausfällt.† Z. B. der Ausdruck $B(Kx_1)x_2$ enthält x_1, wesentlich, aber nicht x_2. Die höchste wesentlich erscheinende Variable in X heisst der *Grad* von X. (Wenn keine Variable wesentlich erscheint, so heisst der Grad 0).

Festsetzung 6. In der Reduktion eines Ausdrucks X auf einen anderen Y heisst eine Variable x_n nicht gestört, wenn 1) X von der Form $X'x_nx_{n+1}\cdots x_{n+p}$ ist, wo X' die Variablen $x_n, x_{n+1}, \cdots, x_{n+p}$ nicht wesentlich enthält, 2) Y von einer ähnlichen Form $Y'x_nx_{n+1}\cdots x_{n+p}$ ist, 3) X' sich auf Y' reduzieren lässt. Sonst heisst eine in X erscheinende Variable gestört.

Festsetzung 7. Ein Ausdruck X *entspricht* einer Folge $\mathfrak{X}$, wenn die fol-

---

*Der leser soll bemerken, dass Ausdrücke wie $X \equiv Y$ und $\vdash X$. Sätze bedeuten. (s. IC).

†Natürlich soll x_m nicht in X selbst fehlen.

or $X_0X_1X_2X_3$ by its equivalent by the corresponding Rule B, C, W, or K, e.g., if $X_0 \equiv B$,* then we have $X_1(X_2X_3)X_4 \cdots X_n$ instead of $X_0X_1X_2X_3X_4 \cdots X_n$. Such a replacement shall be called a *reduction process of the first kind*.[16]

2.) It may be that a constituent of X (i.e., an expression between parentheses appearing in X) can be converted by a reduction process of the first kind. Such conversion is called a *reduction process of the second kind* if a reduction process of the first kind is impossible for the entire expression or any subexpression which either includes the given constituent or is to its left.[17]

Convention 3. An expression X *reduces to* Y if X can be converted to Y by these processes, and this is *in the first sense* if only processes of the first kind are needed and *in the second sense* if processes of the second kind are also needed. That X reduces to Y will be written as $X \dot= Y$.

Convention 4. Let X_m be an expression which contains x_m but no x_n for $n > m$ and is not of the form $X_{m-1}x_m$. We call the infinite sequence of symbols which arises when one puts the variables $x_{m+1}, x_{m+2}, \cdots$ ad infinitum to the right of X_m, the *sequence (Folge)* determined by X_m. The part of this sequence which precedes an x_n, $n > m$, is an expression called a *section* of the sequence. Thus, X_m is itself a section of the sequence it determines.

Convention 5. An expression X contains a variable x_m *essentially (wesentlich)* if for $n >$ all the indexes of variables appearing in X and for $p \geq 0$ the variable x_m is not lost in the reduction of $(Xx_nx_{n+1} \cdots x_{n+p})$.† E.g. the expression $B(Kx_1)x_2$ contains x_1 essentially but not x_2. The highest degree of a variable which appears essentially in X is called the *grade (Grad)* of X.[18] (If no variable appears essentially, the grade is 0.)

Convention 6. In the reduction of an expression X to Y, a variable x_n is *undisturbed (nicht gestört)* if

1) X is of the form $X'x_nx_{n+1} \cdots x_{n+p}$, where X' does not essentially contain the variables $x_n, x_{n+1}, \cdots, x_{n+p}$,

2) Y is of a similar form $Y'x_nx_{n+1} \cdots x_{n+p}$,

3) X' reduces to Y'.

Otherwise a variable appearing in X is disturbed.

Convention 7. An expression X *conforms to (entspricht)* a sequence $\mathfrak{X}$, if the following

---

*The reader must note that expressions like $X \equiv Y$ and $\vdash X$ denote theorems. (Cf. IC).
†Naturally x_m is not itself missing from X.

gende Bedingung erfüllt ist: es gibt ein $p \geqq 0$, so dass der Ausdruck $(Xx_{n+1}x_{n+2}\cdots x_{n+p})$, wo n der Grad von X ist, sich auf einen Abschnitt von $\mathfrak{X}$ reduziert, und zwar so, dass x_{n+p}, wenn $n+p>0$ ist, nicht ausgelassen wird. Wenn x_{n+q} die höchste in dieser Reduktion gestörte Variable ist (bzw. $q=0$, wenn keine nicht in X wesentlich erscheinende Variable gestört wird), so sagen wir dass X der Folge *mit der Ordnung q* entspricht.* Endlich sprechen wir von einem Entsprechen *im ersten bzw. zweiten Sinne*, wenn die Reduktion sich im ersten bzw. zweiten Sinne vollzieht.

Festsetzung 8. Zwei Ausdrücke X und Y heissen *äquivalent* im

1) *ersten Sinne*, wenn sie denselben Grad haben und derselben Folge von lauter Variablen entsprechen,

2) *zweiten Sinne*, wenn sie denselben Grad haben, und derselben Folge von lauter Variablen mit derselben Ordnung entsprechen,

3) *dritten Sinne*, wenn sie denselben Grad haben, und derselben Folge von lauter Variablen in demselben Sinne entsprechen.

4) *vierten Sinne*, wenn sie denselben Grad haben, und derselben Folge von lauter Variablen in demselben Sinne und mit derselben Ordnung entsprechen.

Bemerkung: Die folgenden Sätze haben als Zweck den Beweis, dass, wenn eine Formel der Form $\vdash X = Y$ aus unserem Grundgerüst ableitbar ist, ein gewisser Sinn von Äquivalenz zwischen X und Y besteht. Eine gewisse Art von Übereinstimmung mit Logik und Unabhängigkeit wird dabei für die kombinatorischen Axiome gewährleistet. Dies ist die einzige solche Untersuchung dieser Abhandlung; eine allgemeine Vollständigkeits-, Widerspruchslosigkeits- oder Unabhängigkeitsuntersuchung wird von dieser Abhandlung ausgeschlossen.

Hilfsätze. Das Reduzieren ist seiner Definition nach ein eindeutiger Prozess, also haben wir leicht die folgenden Hilfsätze.

1. Wenn ein Ausdruck sich auf zwei verschiedene Ausdrücke reduziert, so reduziert einer dieser beiden sich auf den anderen.

---

*Man darf hier annehmen dass entweder $p=q$ oder $p=q+1$ ist. Denn nach den Voraussetzungen reduziert $Xx_{n+1}x_{n+2}\cdots x_{n+p}$ sich auf ein $\mathfrak{C}x_{n+q+1}x_{n+q+2}\cdots x_{n+p}$ und zwar so, dass $x_{n+q+1}, x_{n+q+2}, \cdots, x_{n+p}$ dabei ungestört werden. Daher reduziert $Xx_{n+1}x_{n+2}\cdots x_{n+q+1}$ sich auf $\mathfrak{C}x_{n+q+1}$; und dieser ist ein Abschnitt der Folge $\mathfrak{F}$, näm., $\mathfrak{C}x_{n+q+1}x_{n+q+2}\cdots$. Wir wissen ja auch, dass $Xx_{n+1}\cdots x_{n+q}$ sich auf $\mathfrak{C}$ reduziert, aber davon können wir nicht schliessen, dass immer $p=q$ sein kann, weil $\mathfrak{C}$ nicht ein Abschnitt der Folge $\mathfrak{F}$ ist, falls x_{n+q} in der Reduction ausfällt.

condition is satisfied: there is a $p \geq 0$ such that the expression $(Xx_{n+1}x_{n+2}\cdots x_{n+p})$, where n is the grade of X, itself reduces to a section of $\mathfrak{X}$, and this in such a way that x_{n+p}, where $n+p > 0$, does not get deleted. If x_{n+q} is the variable of highest rank disturbed in this reduction ($q = 0$ if no variable not appearing in X essentially becomes disturbed) we say that X conforms to the sequence *with order q* (*mit der Ordnung q*).* Finally, we say that X conforms to the sequence *in the first or second sense* if the reduction is in the first or second sense.

Convention 8. Two expressions X and Y are called *equivalent* in

1) *the first sense* if they have the same grade and conform to the same sequence of variables only,

2) *the second sense* if they have the same grade and conform to the same sequence of variables only with the same order,

3) *the third sense* if they have the same grade and conform to the same sequence of variables only in the same sense,

4) *the fourth sense* if they have the same grade and conform to the same sequence of variables only in the same sense and with the same order.

Remark. The following theorems have as their objective to prove that, if a formula of the form $\vdash X = Y$ is provable from our primitive frame, a certain kind of equivalence holds between X and Y. A certain kind of agreement with logic and independence will thus be guaranteed for the combinatory axioms. This is the only investigation of this treatise; a general investigation of completeness, consistency or independence will be omitted from it.

Lemmas. Reduction is by definition an unambiguous process, so we have easily the following lemmas.

1. If an expression reduces to two different expressions, then one of these expressions reduces to the other.[19]

---

*One may assume here that either $p = q$ or $p = q + 1$. For by the assumptions, $Xx_{n+1}x_{n+2}\cdots x_{n+p}$ reduces to a $\mathfrak{C}x_{n+q+1}x_{n+q+2}\cdots x_{n+p}$ in such a way that $x_{n+q+1}, x_{n+q+2}, \cdots, x_{n+p}$ remain undisturbed. Thus, $Xx_{n+1}x_{n+2}\cdots x_{n+p+1}$ reduces to $\mathfrak{C}x_{n+q+1}$; and this is a section of the sequence $\mathfrak{F}$, namely $\mathfrak{C}x_{n+q+1}x_{n+q+2}\cdots$. We also know that $Xx_{n+1}\cdots x_{n+q}$ reduces to $\mathfrak{C}$, but we cannot conclude from this that we always have $p = q$, because $\mathfrak{C}$ is not a section of the sequence $\mathfrak{F}$ in the case that x_{n+q} drops out of the reduction.

2. Ein Ausdruck kann nie auf zwei verschiedene Kombinationen von Variablen reduziert werden.

3. Ein Ausdruck kann nie zwei verschiedenen Folgen von Variablen entsprechen.

4. Wenn X und Y denselben Grad n haben, und wenn ferner die zwei Ausdrücke $(Xx_{n+1}x_{n+2}\cdots x_{n+p})$ und $(Yx_{n+1}x_{n+2}\cdots x_{n+p})$ sich auf dieselbe Kombination lauter Variablen reduzieren, so sind X und Y äquivalent in dem ersten Sinne.

5. Wenn X und Y denselben Grad n haben, und wenn ferner für jedes p, wofür einer der beiden Ausdrucke $(Xx_{n+1}x_{n+2}\cdots x_{n+p})$ und $(Yx_{n+1}x_{n+2}\cdots x_{n+p})$ auf eine Kombination von lauter Variablen reduziert wird, die beiden sich auf dieselbe Kombination reduzieren, so sind X und Y in dem zweiten Sinne äquivalent.

SATZ 1. *Sind* $\mathfrak{A}_1, \mathfrak{A}_2, \cdots, \mathfrak{A}_N, \mathfrak{B}_1, \mathfrak{B}_2, \cdots, \mathfrak{B}_N, X, Y$, *Kombinationen von Kombinatoren und Variablen* $x_1, x_2, \cdots, x_m$ *derart, dass*

1) *für jedes* i $(i = 1, 2, \cdots N)$ *die beiden Ausdrücke* $\mathfrak{A}_i$ *und* $\mathfrak{B}_i$ *denselben Grad haben, und weiter derselben Folge mit derselben Ordnung und im ersten Sinne entsprechen,*

2) X *einer Folge von lauter Variablen entspricht,*

3) *aus den Voraussetzungen*

(1) $$\vdash \mathfrak{A}_i = \mathfrak{B}_i$$

mit Benutzung nur der Eigenschaften der Gleichheit (I D) folgt, dass

(2) $$\vdash X = Y;$$

dann sind X *und* Y *äquivalent im zweiten Sinne, und zwar, wenn jedes* $\mathfrak{A}_i$ *und jedes* $\mathfrak{B}_i$ *wirklich Kombinatoren enthält, im vierten Sinne.*

Beweis: Es genügt, den Satz für den Fall zu beweisen, dass X aus Y durch eine einzige Ersetzung entsteht, nämlich der Ersetzung eines in X erscheinenden $\mathfrak{A}$ durch seinen Gegenwert $\mathfrak{B}$, oder umgekehrt. Das allgemeinste Y ergibt sich aus X durch eine Reihe von solchen Ersetzungen.

Nach Hp. 2 gibt es ein n, wofür $(Xx_{m+1}x_{m+2}\cdots x_{m+n})$ sich auf eine Kombination von $x_1, x_2, \cdots, x_{m+n}$ reduziert. Ich möchte diese Kombination Z nennen, und die Ausdrücke $(Xx_{m+1}x_{m+2}\cdots x_{m+n})$ bzw. $(Yx_{m+1}x_{m+2}\cdots x_{m+n})$ mit X' und Y' abkürzen. Ich zeige zunächst, dass Y' auf Z reduziert wird, und zwar, wenn die $\mathfrak{A}_i$ und $\mathfrak{B}_i$ alle wirklich Kombinatoren enthalten, in demselben Sinne.

2. An expression can never be reduced to two different combinations of variables.

3. An expression can never conform to two different sequences of variables.

4. If X and Y have the same grade n, and if further the two expressions $(Xx_{n+1}x_{n+2}\cdots x_{n+p})$ and $(Yx_{n+1}x_{n+2}\cdots x_{n+p})$ reduce to the same combination of variables only, then X and Y are equivalent in the first sense.

5. If X and Y have the same grade n, and if further for every p for which the two expressions $(Xx_{n+1}x_{n+2}\cdots x_{n+p})$ and $(Yx_{n+1}x_{n+2}\cdots x_{n+p})$ are reduced to a combination of variables only, both are reduced to the same combination, then X and Y are equivalent in the second sense.

Theorem 1. *If* $\mathfrak{A}_1, \mathfrak{A}_2, \cdots, \mathfrak{A}_N, \mathfrak{B}_1, \mathfrak{B}_2, \cdots, \mathfrak{B}_N, X, Y$ *are combinations of combinators and variables* $x_1, x_2, \cdots, x_m$ *such that*

1) *for every* i ($i = 1, 2, \cdots, N$) *the two expressions* $\mathfrak{A}_i$ *and* $\mathfrak{B}_i$ *have the same grade and further conform to the same sequence with the same order and in the first sense,*

2) X *conforms to a sequence of variables only,*

3) *from the hypotheses*
   (1) $\vdash \mathfrak{A}_i = \mathfrak{B}_i$
   it follows only by the properties of equality (I D) that
   (2) $\vdash X = Y$;

then X *and* Y *are equivalent in the second sense, and, also if every* $\mathfrak{A}_i$ *and every* $\mathfrak{B}_i$ *really contain combinators, in the fourth sense.*

Proof. It suffices to prove the theorem for the case that X is obtained from Y by a single replacement, namely the replacement of $\mathfrak{A}$ by its corresponding $\mathfrak{B}$ or conversely. In the general case, Y can be obtained from X by a sequence of such replacements.

By hypothesis 2, there is an n for which $(Xx_{m+1}x_{m+2}\cdots x_{m+n})$ reduces to a combination of $x_1, x_2, \cdots x_{m+n}$. I will call this combination Z, and I will abbreviate the expressions
$$(Xx_{m+1}x_{m+2}\cdots x_{m+n})$$
and
$$(Yx_{m+1}x_{m+2}\cdots x_{m+n})$$
by X' and Y' respectively. I show first that Y' reduces to Z and, if the $\mathfrak{A}_i$ and $\mathfrak{B}_i$ all really contain combinators, in the same sense.

𝔄 sei der ersetzte Ausdruck in X und 𝔅 sein Gegenwert, so lässt Y' sich von X' nur dadurch unterscheiden, dass in Y' 𝔅 die Stelle von 𝔄 einnimmt. Dann können im Laufe der Reduktion die folgenden drei Möglichkeiten geschehen:

I. Wir kommnen zu einer Form an, worin 𝔄 am Anfang steht, d.h. zu einer Form

(3) $\qquad (𝔄 X_1 X_2 \cdots X_p),$

wo die $X_1, X_2, \cdots, X_p$ Kombinationen von Kombinatoren und Variablen sind.

II. Ein eingeklammerter Teilausdruck, der 𝔄 enthält (bzw. 𝔄 selbst), wird als ein Ganzes durch K ausgestrichen.

III. 𝔄 bleibt innerhalb des Gesamtausdrucks (d.h. nicht am Anfang), bis in der Reduktion durch Prozesse, zweiter Art die Reihe an es kommt, und dann steht es am Anfang eines Teilausdrucks der Form (3), wo $p \geqq 0$ ist.

Diese drei Möglichkeiten sind erschöpfend; weil Reduktion so definiert ist, dass 𝔄 sonst ein untrennbares Ganzes ist. Ich behandle die drei Fälle jetzt besonders.

Fall I. Nach der Voraussetzung dieses Falles reduziert X' sich auf einen Ausdruck X'' der Form (3). Dann reduziert sich Y' durch genau dieselbe Reihe von Reduktionsprozessen auf ein Y'' der Form

(4) $\qquad (𝔅 X_1 X_2 \cdots X_p),$

wo die $X_1, X_2, \cdots, X_p$ dieselben Ausdrücke wie in X'' sind.

Es werde nun angenommen, der Ausdruck $𝔄' \equiv (𝔄 x_{m+1} x_{m+2} \cdots x_{m+p})$ reduziert sich im ersten Sinne auf einen Ausdruck $\mathfrak{C}$. Dann, wenn wir überall in dieser Reduktion $x_{m+1}, x_{m+2}, \cdots x_{m+p}$ durch $X_1, X_2, \cdots X_p$ ersetzen, liefert die so entstehende Folge von Ausdrücken wieder eine Reduktion im ersten Sinne. Daher reduziert sich X'' durch Prozesse erster Art auf ein X''', welches entsteht, wenn man in $\mathfrak{C}$ die betreffenden Einsetzungen macht. Eine ähnliche Bemerkung bezieht sich auf Y''.

Nach Hp. 1 entsprechen 𝔄 und 𝔅 beide derselben Folge 𝔉. r sei die Ordnunug, womit 𝔄 dem 𝔉 entspricht. Dann zeige ich, dass $r \leqq p$ ist. In der Tat nehmen wir das Gegenteil an. Dann reduziert der Ausdruck $(𝔄' x_{m+p+1} x_{m+p+2} \cdots x_{m+r+1})$ sich auf einen Abschnitt von 𝔉 und zwar so, dass x_{m+p+1} gestört wird.* Unter der durch diese Reduktion erzeugten Reihe von Ausdrücken gibt es ein $(\mathfrak{C} x_{m+p+1} x_{m+p+2} \cdots x_{m+r+1})$ derart, dass die Reduktion sich bis auf diesen Ausdruck ohne Störung von $x_{m+p+1} \cdots x_{m+r+1}$ erstrickt,

---

*S. Festsetzung 6, Anmerkung.

Let $\mathfrak{A}$ be the given expression in X and $\mathfrak{B}$ its correspondent, so that the difference between Y' and X' is that in Y' $\mathfrak{B}$ stands in place of $\mathfrak{A}$. Then there are three possibilities that can appear in the course of the reduction:

I. We come to a form where $\mathfrak{A}$ stands at the start, i.e. to a form
   (3) $(\mathfrak{A} X_1 X_2 \cdots X_p)$,

   where $X_1, X_2, \cdots, X_p$ are combinations of combinators and variables.

II. A subexpression enclosed in parentheses that contains $\mathfrak{A}$ (or perhaps is $\mathfrak{A}$ itself) is erased as a whole by K.

III. $\mathfrak{A}$ remains inside the entire expression (i.e., not at the beginning) until a point comes in the reduction process of the second kind that it stands at the beginning of a subexpression of the form (3) where $p \geq 0$.

These three possibilities are exhaustive because reduction is so defined that $\mathfrak{A}$ is an otherwise unbreakable whole. I will now take up the three cases separately.

Case I. By the assumption of this case, X' reduces to an expression X'' of the form (3). Then Y' reduces by the same sequence of reduction processes to a Y''' of the form

(4) $(\mathfrak{B} X_1 X_2 \cdots X_p)$,

where the $X_1, X_2, \cdots, X_p$ are the same expressions as in X''.

It will now be assumed that the expression $\mathfrak{A}' \equiv (\mathfrak{A} x_{m+1} x_{m+2} \cdots x_{m+p})$ reduces in the first sense to an expression $\mathfrak{C}$. Then, if we replace in this reduction $x_{m+1}, x_{m+2}, \cdots, x_{m+p}$ by $X_1, X_2, \cdots, X_p$ respectively, the resulting sequence of expressions becomes a reduction in the first sense. Hence X'' reduces by a process of the first kind to an X''', which results when one makes in $\mathfrak{C}$ the corresponding substitutions. A similar remark holds also for Y''.

By hypothesis 1, $\mathfrak{A}$ and $\mathfrak{B}$ conform to the same sequence $\mathfrak{F}$. Let r be the order in which $\mathfrak{A}$ conforms to $\mathfrak{F}$. Then I will show that $r \leq p$. In fact, let us assume the opposite. Then the expression $(\mathfrak{A}' x_{m+p+1} x_{m+p+2} \cdots x_{m+r+1})$ reduces to a section of $\mathfrak{F}$ in such a way that x_{m+p+1} is disturbed.* In the sequence of expressions generated by this reduction there is a $(\mathfrak{C} x_{m+p+1} x_{m+p+2} \cdots x_{m+r+1})$ such that the reduction reaches this expression without disturbing $x_{m+p+1}, \cdots, x_{m+r+1}$, while

---
*Cf. Convention 6, Remark.

während im nächsten Schritte der Reduktion x_{m+p+1} gestört wird. Also muss $\mathfrak{C}$ von der Form
(5) $\qquad (X_0'' X_1'' X_2'' \cdots X_q'')$
sein, wo entweder 1) $X_0'' B$ oder C ist und $q < 3$ ist, oder 2) $X_0'' W$ oder K ist und $q < 2$ ist. Nach Festsetzung 6 reduziert $\mathfrak{A}'$ sich auf dieses $\mathfrak{C}$. Dann reduziert sich X'' nach dem vorigen Absatz auf ein X''' derselben Form (5). Aber in der weiteren Reduktion eines solchen X''' könnte der Kombinator X_0'' nie verschwinden, was der Voraussetzung, dass X' sich auf Z reduziert, widerspricht.

Also gilt $r \leqq p$. Dann reduzieren sich die beiden Ausdrücke $(\mathfrak{A}' x_{m+p+1})$ und $(\mathfrak{B}' x_{m+p+1})$ auf einen Abschnitt $(\mathfrak{C} x_{m+p+1})$ von $\mathfrak{F}$, und zwar so, dass x_{m+p+1} ungestört wird. Daher reduzieren sich $\mathfrak{A}'$ und $\mathfrak{B}'$ beide auf dasselbe $\mathfrak{C}$ (Festsetzung 6). Diese Reduktion geschieht weiterhin im ersten Sinne. Nach dem vorletzten Absatz reduzieren sich dann X'' und Y'' auf ein gemeinsames X''', und zwar im ersten Sinne. Weil X' auf Z reduziert wird, so reduziert sich X''', und also Y' auf Z. Weil die einzigen Reduktionsprozesse, die in den Reduktionen von X' und Y' verschieden sind, zu der ersten Art gehören, so reduzieren X' und Y' sich auf Z in demselben Sinne.

Fall II. Durch eine Reihe von Reduktionsprozessen reduziert X' sich auf einen Ausdruck X'', der einen Teilausdruck der Form $(K X_1 X_2 \cdots X_p)$ enthält, wo $\mathfrak{A}$ in X_2 enthalten ist, und zwar so, dass beim nächsten Schritte die Reduktion auf einen X'' führt, der sich vom X'' nur dadurch unterscheidet, dass der obige Teilausdruck durch $(X X_3 \cdots X_p)$ ersetzt ist. Genau dieselbe Reihe von Prozessen reduziert Y' auf einen Ausdruck Y'', der sich von X'' nur darin unterscheidet, dass $\mathfrak{A}$ die Stelle von $\mathfrak{B}$ einnimmt. Beim nächsten Schritte, der derselbe Prozess wie im vorigen Falle ist, kommen wir wieder auf X'''. Daher reduzieren X' und Y' sich durch dieselbe Reihe von reduktionprozessen auf denselben Ausdruck. Infolgedessen reduzieren sie sich endlich auf dieselben Kombination, und zwar weil die beiden Reihen von Prozessen dieselben sind, in demselben Sinne.

Fall III. Nach der Voraussetzung reduziert X' sich auf einen Ausdruck X'', der einen Teilausdruck der Form (3) enthält, und zwar so, dass die weitere Reduktion von X'' durch die Reduktion dieses Teilausdrucks fortgesetzt wird. Dann reduziert Y' sich auf ein Y'' welches sich von X'' nur darin unterscheidet, dass der Ausdruck (4) anstatt (3) erscheint.

Weil die Bedingungen von Fall I für diese Teilausdrücke (3) und (4) erfüllt sind, so reduzieren diese Teilausdrücke sich auf dieselben Kombinationen. Weil X'' und Y'' sonst identisch sind, so reduzieren X'' und Y'',

in the next step of the reduction x_{m+p+1} is disturbed. Then $\mathfrak{C}$ must have the form

(5) $(X_0'' X_1'' X_2'' \cdots X_q'')$

where either 1) X_0'' is B or C and $q < 3$, or 2) X_0'' is W or K and $q < 2$. By Convention 6 $\mathfrak{A}'$ reduces to this $\mathfrak{C}$. Then by the previous paragraph X'' reduces to an X''' of the same form (5). But in the further reduction of such an X''' the combinator X_0'' can never disappear, which contradicts the assumption that X' reduces to Z.

Thus, $r \leq p$. Then both expressions $(\mathfrak{A}' x_{m+p+1})$ and $(\mathfrak{B}' x_{m+p+1})$ reduce to a section $(\mathfrak{C} x_{m+p+1})$ of $\mathfrak{F}$ in such a way that x_{m+p+1} is undisturbed. Hence, $\mathfrak{A}'$ and $\mathfrak{B}'$ reduce to this same $\mathfrak{C}$ (Convention 6). Furthermore, this reduction occurs in the first sense. By the paragraph before last, X'' and Y'' reduce to a common X''', also in the first sense. Because X' reduces to Z, so X''', and hence Y', reduces to Z. Because the reduction processes, which differ in the reductions from X' and Y', belong to the first kind, X' and Y' reduce to Z in the same sense.

Case II. By a series of reduction processes X' reduces to an expression X'' which contains a subexpression of the form $(K X_1 X_2 \cdots X_p)$ where $\mathfrak{A}$ is contained in X_2 in such a way that the next step of the reduction leads to an X''' which differs from X'' only in that the above subexpression is replaced by $(X_1 X_3 \cdots X_p)$. Exactly the same series of processes reduces Y' to an expression Y'' which differs from X'' only in that the latter has $\mathfrak{A}$ in place of $\mathfrak{B}$. In the next step of the same process, as in the previous case, we come to X'''. Thus, X' and Y' are reduced by the same series of reduction processes to the same expression. Thus, they finally reduce to the same combination and, because the series of processes are the same, in the same sense.

Case III. By the assumption, X' reduces to an expression X'' which contains a subexpression of the form (3) in such a way that the further reduction of X'' is continued through a reduction of this subexpression. Then Y' reduces to a Y'' which differs from X'' only in that the expression (4) appears instead of (3).

Because the assumptions of Case I are fulfilled for these subexpressions (3) and (4), these subexpressions reduce to the same combinations. Because X'' and Y'' are otherwise identical, X'' and Y''',

und daher auch X' und Y' sich auf denselben Ausdruck. Infolgedessen werden X' und Y' auf dieselbe Kombination von Variablen reduziert.

Wenn $\mathfrak{A}$ und $\mathfrak{B}$ wirklich Kombinatoren enthalten, so sind Reduktionsprozesse zweiter Art in den beiden Fällen erforderlich. Deshalb werden sie auf diese Kombination in demselben Sinne reduziert.

Es ist nun bewiesen, dass X' und Y' sich auf dasselbe Z reduzieren. Daraus folgt zunächst, dass X und Y denselben Grad haben; denn jede Variable, die in der Reduktion von X' verschwindet, verschwindet auch in der Reduktion von Y' und umgekehrt. Dieser Grad sei dann μ. Setzen wir in den obigen Beweis $x_{\mu+j}$ statt x_{m+j} ein, so folgt, dass die neuen X' und Y' auch auf eine gemeinsame Kombination lauter Variablen reduziert werden, wenn nur eines von den beiden sich auf eine solche Kombination reduziert. Also entsprechen X und Y derselben Folge mit derselben Ordnung (Hilfsatz 5), und auch, wenn die $\mathfrak{A}$ und $\mathfrak{B}$ wirklich Kombinatoren enthalten, in dem selben Sinne, w.z.b.w.

SATZ 2. *Sind* $\mathfrak{A}_1, \mathfrak{A}_2, \cdots, \mathfrak{A}_N, \mathfrak{B}_1 \mathfrak{B}_2, \cdots, \mathfrak{B}_N, X, Y$, *Kombinationen von Kombinatoren und Variablen* $x_0, x_1, \cdots, x_m$ *derart, dass die Bedingungen von Satz 1 erfüllt sind, ausser dass in Hp. 3 bei der Ableitung von (2) aus (1) auch Benutzung von den Regeln B, C, W und K erlaubt wird; dann sind* X *und* Y *im zweiten Sinne äquivalent.*

Beweis: Wir können von X zu Y durch eine Reihe von Schritten übergehen, wovon jeder daraus besteht, dass wir entweder eine einzige Ersetzung aus den Formeln (1) machen, oder auch eine Regel B, C, W oder K einmal anwenden. Weiter dürfen wir unter einer solchen Anwendung den folgenden Prozess verstehen: zunächst setzen wir in einer Regel (B, C, W oder K) für die X, Y (und Z, wenn es erscheint) besondere Ausdrücke ein, sodass eine Formel $\mathfrak{A} = \mathfrak{B}$ entsteht, und dann machen wir in einem schon aus X abgeleiteten Ausdruck eine Ersetzung von $\mathfrak{A}$ durch $\mathfrak{B}$ oder umgekehrt.

Jetzt betrachten wir alle die Formeln, die in diese Weise aus allen den im Uebergang von X zu Y benutzten Anwendungen der betreffenden Regeln entstehen. Fügen wir diese Formeln zu den Formeln (1) hinzu. Dann sind alle die Bedingungen von Satz 1 für die erweiterte $\mathfrak{A}_1, \mathfrak{A}_2, \cdots \mathfrak{A}_N, \mathfrak{A}_{n+1}, \cdots, \mathfrak{A}_M, \mathfrak{B}_1, \mathfrak{B}_2, \cdots, \mathfrak{B}_N, \mathfrak{B}_{N+1}, \cdots, \mathfrak{B}_M, X, Y$ erfüllt. Also folgt der Satz aus Satz 1.

Es soll bemerkt werden, dass die Nebenbedingung für den strengen Satz 1, nämlich dass alle die $\mathfrak{A}_i$ und $\mathfrak{B}_i$ wirklich Kombinatoren enthalten, für die neue $\mathfrak{A}_{N+j}$ oder $\mathfrak{B}_{N+j}$ versagen mag, sogar wenn sie für die ursprüngliche $\mathfrak{A}_i$ und $\mathfrak{B}_i$ erfüllt ist.

and hence also X' and Y', reduce to the same expression. Hence, X' and Y' reduce to the same combination of variables.

If $\mathfrak{A}$ and $\mathfrak{B}$ really contain combinators, reduction processes of the second kind are necessary. Hence, they reduce to these combinations in the same sense.

It has now been proved that X' and Y' reduce to the same Z. Next it follows that X and Y have the same grade; for every variable which disappears in the reduction of X' also disappears in the reduction of Y', and conversely. Let this grade be μ. If we put $x_{\mu+j}$ for x_{m+j} in the above proof, it follows that the new X' and Y' also reduce to a common combination of variables only, when only one of the two reduces to such a combination. Thus, X and Y conform to the same sequence with the same order (Lemma 5), and also if $\mathfrak{A}$ and $\mathfrak{B}$ really contain combinators, in the same sense, q.e.d.

Theorem 2. *If $\mathfrak{A}_1, \mathfrak{A}_2, \cdots, \mathfrak{A}_N, \mathfrak{B}_1, \mathfrak{B}_2, \cdots, \mathfrak{B}_N, X, Y$ are combinations of combinators and variables $x_0, x_1, \cdots, x_m$ so that the hypotheses of Theorem 1 are satisfied except that in hypothesis 3 for the derivation of (2) from (1) the uses of rules B, C, W, and K are also allowed, then X and Y are equivalent in the second sense.*

Proof. We can pass from X to Y by a series of steps, each of which consists either of a single replacement by formula (1) or else uses Rule B, C, W, or K once. Furthermore, we can understand by one such use the following process: first, for X, Y (and Z if it appears) in a rule (B, C, W, or K), we substitute certain expressions in such a way that a formula $\mathfrak{A} = \mathfrak{B}$ results, and then we make in an expression already derived in X a replacement of $\mathfrak{A}$ by $\mathfrak{B}$ or conversely.

Now we consider all the formulas which arise in this way from all applications of the rules used in the passage from X to Y. We add these formulas to formula (1). Then all the assumptions of Theorem 1 hold for the expanded $\mathfrak{A}_1, \mathfrak{A}_2, \cdots, \mathfrak{A}_N, \mathfrak{A}_{N+1}, \cdots, \mathfrak{A}_M, \mathfrak{B}_1, \mathfrak{B}_2, \cdots, \mathfrak{B}_N, \mathfrak{B}_{N+1}, \cdots, \mathfrak{B}_M, X, Y$. The theorem then follows by Theorem 1.

It should be remarked that the additional postulate for the strict Theorem 1, namely that all the $\mathfrak{A}_i$ and $\mathfrak{B}_i$ really contain combinators, may be false for the new $\mathfrak{A}_{N+j}$ or $\mathfrak{B}_{N+j}$ even if it holds for the original $\mathfrak{A}_i$ and $\mathfrak{B}_i$.

SATZ 3. *Wenn* $\mathfrak{A}_1, \mathfrak{A}_2, \cdots, \mathfrak{A}_N, \mathfrak{B}_1, \mathfrak{B}_2, \cdots, \mathfrak{B}_N, X, Y$ *Kombinatoren sind, die die Hypothesen von Satz 2 erfüllen; dann sind X und Y äquivalent im vierter Sinne.*

Beweie: Die $\mathfrak{A}_i$ und $\mathfrak{B}_i$, die sowohl in den ursprünglichen Formeln (1), als auch in denen, die dazu durch die Prozesse des Beweises von Satz 2 hinzugefügt werden, erscheinen, sind Kombinatoren und enthalten daher Kombinatoren. Also folgt der Satz aus Satz 1.

SATZ 4. *Sind X, Y Kombinatoren, wofür*

1) es folgt aus den transmutativen Axiomen mit Benutzung der Regeln B, C, W, K und den Eigenschaften der Gleichheit, dass $\vdash X = Y$,

2) mindestens einer der beiden einer Folge von lauter Variablen entspricht; dann sind X und Y äquivalent im vierten Sinne.

Beweis: folgt aus Satz 3, weil die betreffenden Axiome die Bedingungen der Formeln (1) erfüllen.

SATZ 5. *Wenn* $\mathfrak{A}_1, \mathfrak{A}_2, \cdots, \mathfrak{A}_N, \mathfrak{B}_1, \mathfrak{B}_2, \cdots, \mathfrak{B}_N, X, Y$ *Kombinationen von Variablen und Kombinatoren sind, derart, dass*

1) für jedes i $(i = 1, 2, \cdots, N)$ $\mathfrak{A}_i$ *und* $\mathfrak{B}_i$ *denselben Grad haben und weiter derselben Folge mit derselben Ordnung entsprechen,*

2) X einer Folge von lauter Variblen entspricht,

3) aus den Formeln

(1) $$\vdash \mathfrak{A}_i = \mathfrak{B}_i \qquad (1 = 1, 2, \cdots, N)$$

mit Benutzung der Regeln B, C, W, K und der Eigenschaften der Gleichheit folgt, dass

(2) $$\vdash X = Y;$$

dann sind X und Y äquivalent im zweiten Sinne.

Beweis: Zunächst sehen wir sofort, dass der Satz, wenn er für den Fall bewiesen ist, dass im Hp. 3) die Benutzung nur von den Eigenschaften der Gleichheit erlaubt ist, im allgemeinen durch das Verfahren, das ich in dem Beweis von Satz 2 benutzt habe, bewiesen werden kann. Es genügt daher, den Satz für jenen Fall zu beweisen.

Der Beweis verläuft nun wie der von Satz 1. Wir setzen ohne Beschränkung der Allgemeinheit voraus, dass Y sich aus X durch eine einzige Einsetzung, die von $\mathfrak{A}$ statt $\mathfrak{B}_2$ ergibt. Wir definieren X', Y' und Z wie dort, und schliessen, wie folgt, dass Y' auf Z reduziert wird. Wir unterscheiden dieselben drei Fälle wie im Satz 1.

Theorem 3. *If $\mathfrak{A}_1, \mathfrak{A}_2, \cdots, \mathfrak{A}_N, \mathfrak{B}_1, \mathfrak{B}_2, \cdots, \mathfrak{B}_N, X, Y$ are combinators which satisfy the hypotheses of Theorem 2, then X and Y are equivalent in the fourth sense.*

Proof. The $\mathfrak{A}_i$ and $\mathfrak{B}_i$, which appear both in the original formula (1) and also in those which are added by the process of the proof of Theorem 2 are combinators and thus contain combinators. Thus the theorem follows by Theorem 1.

Theorem 4. *Let X, Y be combinators for which*

1) *it follows from the transmutative axioms with the use of Rules B, C, W, and K and properties of equality that $\vdash X = Y$,*

2) *at least one of the two conforms to a sequence of variables only,*

then X and Y are equivalent in the fourth sense.

Proof. This follows from Theorem 3 because the axioms in question satisfy the conditions of Theorem 1.

Theorem 5. *If $\mathfrak{A}_1, \mathfrak{A}_2, \cdots, \mathfrak{A}_N, \mathfrak{B}_1, \mathfrak{B}_2, \cdots, \mathfrak{B}_N, X, Y$ are combinations of variables and combinators such that*

1) *for every i $(i = 1, 2, \cdots, N)$ $\mathfrak{A}_i$ and $\mathfrak{B}_i$ have the same grade and conform to the same sequences with the same order,*

2) *X conforms to a sequence of variables only,*

3) *from the formulas*

(1) $\quad \vdash \mathfrak{A}_i = \mathfrak{B}_i \qquad\qquad (i = 1, 2, \cdots, N)$

it follows by the rules B, C, K, and W and the properties of equality that

(2) $\quad \vdash X = Y$;

then X and Y are equivalent in the second sense.

Proof. First, we can easily see that the theorem can be proved in full generality using the procedure that I used in the proof of Theorem 2, if it can be proved for the case that in hypothesis 3) only the use of properties of equality is allowed. It is therefore enough to prove the theorem for that case.

The proof now proceeds like that of Theorem 1. We assume without loss of generality that Y comes from X by a single replacement of $\mathfrak{A}$ by $\mathfrak{B}$. We define X', Y', and Z as there and conclude as follows that Y' reduces to Z. We differentiate the same three cases as in Theorem 1.

Fall I. X' und Y' reduzieren sich auf X'' bzw. Y'' von der Form (3) bzw. (4) .

Es sei nun angenommen, der Ausdruck $\mathfrak{A}'$ (definiert wie im Satz 1) reduziert sich auf einen Ausdruck $\mathfrak{C}$; dann, wenn wir überall in dieser Reduktion $x_{m+1}, x_{m+2}, \cdots, x_{m+p}$ durch $X_1, X_2, \cdots, X_p$ ersetzen, so schaffen wir eine Reihe von Ausdrücken, die, obgleich sie nicht immer eine Reduktion liefern müssen, doch nach Satz 2 (für $N = 0$) immer zueinander im zweiten Sinne äquivalent sind. Infolgedessen muss X'' und daher auch X' mit einem X''', das aus $\mathfrak{C}$ durch die erwähnte Einsetzung entsteht, im zweiten Sinne äquivalent sein.

Nach Hp. 1 entsprechen $\mathfrak{A}$ und $\mathfrak{B}$ derselben Folge $\mathfrak{F}$. r sei die Ordnung, womit $\mathfrak{A}$ dem $\mathfrak{F}$ entspricht Dann gilt $r \leq p$. In der Tat sei angenommen, dass $r > p$ ist. Dann folgt, genau wie in Satz 1, dass $\mathfrak{A}'$ auf ein $\mathfrak{C}$ der Form (5) reduziert wird. Daher ist X'', nach dem vorigen Absatz, mit einem X''' der Form (5) im zweiten Sinne aquivalent. Dies ist aber unmöglich, weil X' und daher X'', einer mit Z anfangenden Folge lauter Variablen mit der Ordnung 0 entspricht, während X''' keiner Folge lauter Variablen mit der Ordnung 0 entsprechen kann.

Es folgt dann, wie im Satz 1, dass $\mathfrak{A}'$ und $\mathfrak{B}'$ sich auf dasselbe $\mathfrak{C}$ reduzieren. Daher sind X und Y nach dem vorletzten Absatz mit demselben X''' im zweiten Sinne äquivalent. Aber nach der Voraussetzung reduziert X' sich auf Z. Daher reduziert sich auch Y' auf Z.

Der Rest des Beweises verläuft genau wie im Satz 1.

SATZ 6, *Sind X und Y Kombinatoren, wofür*

1) *mindestens einer der beiden einer Folge von lauter Variablen entspricht,*

2) *aus den transmutativen und kommutativen Axiomen folgt, dass* $\vdash X = Y$;

dann sind X und Y äquivalent im zweiten Sinne.

Beweis: Folgt aus Satz 5, weil die betreffenden Axiome die Bedingungen der Formel (1) erfüllen.

SATZ 7. *Ax. I_2 ist nicht aus den übrigen kombinatorischen Axiomen mit Benutzung der Regeln B, C, K, W und den Eigenschaften der Gleichheit ableitbar.*

Beweis: Folgt aus Satz 6, weil die zwei Kombinatoren, die in Ax. I_2 auf den beiden Seiten des Zeichens = stehen, im dritten, aber nicht im zweiten Sinne äquivalent sind.

Case I. X' and Y' reduce to X'' respectively Y'' of the form (3) respectively (4).

Let it now be assumed that the expression $\mathfrak{A}'$ (defined as in Theorem 1) reduces to an expression $\mathfrak{C}$; then when we replace $x_{m+1}, x_{m+2}, \cdots, x_{m+p}$ in this reduction by $X_1, X_2, \cdots, X_p$, we create a series of expressions, which although they don't always deliver a reduction, are still always equivalent in the second sense, by Theorem 2 (for $N = 0$). Consequently X'', and therefore also X', are equivalent in the second sense to an X''', which also results from $\mathfrak{C}$ by the above mentioned replacement.

By hypothesis 3 $\mathfrak{A}$ and $\mathfrak{B}$ conform to the same sequence $\mathfrak{F}$. Let r be the order in which $\mathfrak{A}$ conforms to $\mathfrak{F}$. Then $r \leq p$. For, assume that $r > p$. Then it follows exactly as in Theorem 1, that $\mathfrak{A}'$ reduces to a $\mathfrak{C}$ of the form (5). Hence X'', by the previous paragraph, is equivalent to an X''' of the form (5) in the second sense. But this is impossible, because X', and therefore X'', conforms with order 0 to a sequence of variables only starting with Z, whereas X''' cannot conform with order 0 to any sequence of variables only.

It follows therefore, as in Theorem 1, that $\mathfrak{A}'$ and $\mathfrak{B}'$ reduce to the same $\mathfrak{C}$. Hence, X and Y by the paragraph before last are equivalent with the same X''' in the second sense. But by hypothesis X' reduces to Z. Hence also Y' reduces to Z.

The rest of the proof runs exactly as in Theorem 1.

Theorem 6. *Let X and Y be combinators for which*

1) *at least one of the two conforms to a sequence of variables only,*

2) *by the transmutative and commutative axioms it follows that $\vdash X = Y$;*

then X and Y are equivalent in the second sense.

Proof. Follows from Theorem 5, because the axioms in question satisfy the conditions of Formula (1) in that theorem.

Theorem 7. *Ax. I_2 is not derivable from the rest of the combinatory axioms using the rules B, C, K, and W and the properties of equality.*

Proof. Follows from Theorem 6, because the two combinators, which appear in Ax. I_2 on both sides of the $=$ sign, are equivalent in the third, not in the second sense.

SATZ 8. *Wenn wir in den Hypothesen von Sätzen 1-3 und 5 die folgenden Änderungen machen*:

1) $\mathfrak{A}_i$ und $\mathfrak{B}_i$ brauchen nicht dieselbe Ordnung (inbezug auf ihr Entsprechen einer gemeinsamen Folge) zu haben,

2) nicht nur X, sondern auch Y einer Folge von lauter Variablen entspricht;

dann folgen die Schlüsse dieser Sätze, wenn wir darin den vierten Sinn durch den dritten, und den zweiten Sinn durch den ersten ersetzen.

Beweis: Die einzigen Stellen in den Beweisen der betr. Sätze, wo wir die Voraussetzung über die Ordnung von den $\mathfrak{A}_i$ und $\mathfrak{B}_i$ benutzt haben, sind im Fall I unter den Sätzen 1 und 5, und zwar wird sie da nur benutzt, um zu beweisen, dass $\mathfrak{A}'$ und $\mathfrak{B}'$ sich auf ein gemeinsames $\mathfrak{C}$ reduzieren.

Diesen Schluss können wir auch im vorliegenden Falle erreichen. Es folgt ohne Benutzung der betr. Voraussetzung, dass entweder $\mathfrak{A}'$ und $\mathfrak{B}'$ sich auf ein gemeinsames $\mathfrak{C}$ reduzieren, oder einer der beiden auf einen Ausdruck der Form (5) reduziert wird. n sei nun so gewählt, dass nicht nur X', sondern auch Y' sich auf eine Kombination von lauter Variablen reduziert. Dies ist möglich nach Hp. 2 dieses Satzes. Dann folgt durch das Argument des dritten Absatzes des Falles 1 in den Sätzen 1 und 5, dass weder $\mathfrak{A}'$ noch $\mathfrak{B}'$ sich auf einen Ausdruck der Form (5) reduzieren lässt. Daher müssen sie sich auf einen gemeinsamen $\mathfrak{C}$ reduzieren.

Diese Änderung des n stört aber nichts in den Beweisen der betr. Sätze, ausser dass wir jetzt nicht schliessen können, dass X und Y dieselbe Ordnung haben. Also haben wir einen wirklichen Beweis, wenn wir die ganzen Beweise hindurch die Ersetzungen vom Schlusse dieses Satzes machen. Damit wird die Behauptung bewiesen.

SATZ 9. *Wenn X und Y Kombinatoren sind, wofür*

1) sowohl X wie auch Y einer Folge von lauter Variablen entspricht,

2) aus den transmutativen Axiomen und Ax. I_2 mit Benutzung der Eigenschaften der Identität und Regeln B, C, K, W folgt, dass $\vdash X = Y$; dann sind X und Y im dritten Sinne äquivalent.

Beweis: Folgt aus Sätzen 3 und 8.

SATZ 10. *Dir kommutativen Axiome sind nicht Folgerungen aus den anderen kombinatorischen Axiomen.*

Beweis: Die Kombinatoren, die in diesen Axiomen auf den beiden Seiten des Zeichens = stehen, sind nicht im dritten Sinne äquivalent. Daher folgt der Satz aus Satz 9.

Theorem 8. *If in the hypotheses of Theorems 1-3 and 5 we make the following changes:*

1) *$\mathfrak{A}_i$ and $\mathfrak{B}_i$ need not have the same order (with respect to the common sequence to which they conform),*

2) *not only X, but also Y conforms to a sequence of variables only;*

then the conclusions of these theorems hold, where we replace the fourth sense by the third, and the second sense by the first.

Proof. The only places in the proofs of the relevant theorems where we used the hypothesis on the order of $\mathfrak{A}_i$ and $\mathfrak{B}_i$ are in case I of Theorems 1 and 5, and there it was only used to prove that $\mathfrak{A}'$ and $\mathfrak{B}'$ reduce to a common $\mathfrak{C}$.

We can also get this conclusion in the present cases. It follows without using the hypothesis in question that either $\mathfrak{A}'$ and $\mathfrak{B}'$ reduce to a common $\mathfrak{C}$, or one of them reduces to an expression of the form (5). Choose n[20] now so that not only X', but also Y', reduces to a combination of variables only. This is possible by hypothesis 2 of this theorem. Then it follows by the argument of the third paragraph of Case 1 in Theorems 1 and 5 that neither $\mathfrak{A}'$ nor $\mathfrak{B}'$ can be reduced to an expression of the form (5). Therefore, they must reduce to a common $\mathfrak{C}$.

This change of n does not affect anything in the proofs of the relevant theorems except that we now cannot conclude that X and Y have the same order. Hence we have a real proof when we make the replacements of the conclusions of these theorems throughout all the proofs. With this the assertion is proven.

Theorem 9. *Let X and Y be combinators for which*

1) *both X and Y conform to a sequence of variables only,*

2) *by the transmutative axioms and Ax. I_2 using the properties of identity and rules B, C, K, and W it follows that $\vdash X = Y$;*

then X and Y are equivalent in the third sense.

Proof. Follows from Theorems 3 and 8.

Theorem 10. *The commutative axioms are not consequences of the other combinatory axioms.*

Proof. The combinators, which appear in these axioms on both sides of the $=$ sign, are not equivalent in the third sense. Hence the theorem follows from Theorem 9.

SATZ 11. *Sind X und Y Kombinationen von Variablen und Kombinatoren derart, dass*

1) sowohl X wie auch Y einer Folge lauter Variablen entspricht,

2) aus den kombinatorischen Axiomen überhaupt mit Benutzung der Eigenschaften der Gleichheit und der Regeln B, C, K, W folgt dass $\vdash X = Y$;

dann haben X und Y denselben Grad, und sie entsprechen derselben Folge.

Beweis: Nach den Sätzen 5 und 8 sind X und Y im ersten Sinne äquivalent. Daher folgt der Satz gleich aus der Definition der Äquivalenz.

SATZ 12. *Sind X und Y Kombinationen lauter Variablen, wofür die Hp. 2 von Satz 11 erfüllt ist, so sind X und Y identisch.*

Beweis: Nach Satz 11 entsprechen X und Y derselben Folge; dies kann nur geschehen, wenn sie Abschnitte derselben Folge sind. Weiter haben sie nach Satz 11 denselben Grad; daraus folgt, dass sie genau derselbe Abschnitt sind.

Festsetzung 9. Ein Kombinator X stellt eine Kombination Y der Variablen $x_1, x_2, \cdots, x_n$ dann und nur dann dar, wenn aus den kombinatorischen Axiomen mit Benutzung der Eigenschaften der Gleichheit und der Regeln B, C, W, K folgt, dass

$$\vdash X x_1 x_2 \cdots x_n = Y.$$

SATZ 13. *Wenn ein Kombinator eine Kombination von $x_1, x_2, \cdots x_n$ darstellt, so stellt er nur eine dar.*

Beweis: Folgt gleich aus Satz 12.

§2. *Normale Kornbinationen und Folgen.*

Festsetzung 1. Unter einer *normalen Kombination* von $X_0, X_1, X_2, \cdots, X_n$ verstehen wir einen Ausdruck der Form

$$(X_0 Y_1 Y_2 \cdots Y_n),$$

wo jedes Y_i eine Kombination von $X_1, X_2, \cdots, X_n$ ist.

Festsetzung 2. Hiernach wird zuweilen auch das Zeichen x_0 als Variable gebraucht.[*]

---

[*]In der inhaltlichen Anwendung der vorliegenden Theorie wird im allgemeinen eine Funktion (wie ϕ in II A 3), die Stelle von x_0 einnehmen. Die Variable x_0 wird hiernach im allgemeinen nur für normalen Folgen usw. benutzt.

Theorem 11. *If X and Y are combinations of variables and combinators such that*

1) *both X and Y conform to a sequence of variables only,*

2) *by the combinatory axioms together with use of the properties of equality and rules B, C, K, and W it follows that $\vdash X = Y$;*

then X and Y have the same grade and conform to the same sequence.

Proof. By Theorems 5 and 8, X and Y are equivalent in the first sense. Therefore the theorem follows immediately from the definition of equivalence.

Theorem 12. *If X and Y are combinations of variables only such that hypothesis 2 of Theorem 11 is satisfied, then X and Y are identical.*

Proof. By Theorem 11, X and Y conform to the same sequence; this can only occur when they are sections of the same sequence. Furthermore, by Theorem 11 they have the same grade; hence it follows that they are exactly the same section.

Convention 9. A combinator X *represents* a combination Y of variables $x_1, x_2, \cdots, x_n$ if and only if, it follows from the combinatory axioms with use of the properties of equality and the rules B, C, W, and K that

$$\vdash X x_1 x_2 \cdots x_n = Y.$$

Theorem 13. *If a combinator represents a combination of $x_1, x_2, \cdots, x_n$, then it represents only one.*

Proof. Follows immediately from Theorem 12.

§2. *Normal combinations and sequences.*

Convention 1. By a *normal combination* of $X_0, X_1, X_2, \cdots, X_n$ we understand an expression of the form $(X_0 Y_1 Y_2 \cdots Y_n)$ where every Y_i is a combination of $X_1, X_2, \cdots, X_n$.

Convention 2. From now on, the symbol x_0 will sometimes be used as a variable.*

---

*In the contensive application of the present theory, a function (like ϕ in II A 3) will take the place of x_0. In what follows, the variable x_0 will in general only be used for normal sequences, etc.

Festsetzung 3. Unter einer *normalen Folge* (von Variablen) verstehen wir eine Folge, die durch eine normale Kombination von $x_0, x_1, x_2, \cdots, x_n$ bestimmt ist (§1, Festsetzung 4), wo n irgendeine ganze Zahl > 0 Solche normalen Folgen werden hiernach mit griechischen Buchstaben bezeichnet.

Festsetzung 4. Unter dem Produkt $(\eta \cdot \zeta)$ von zwei normalen Folgen η und ζ verstehen wir die folgendermassen bestimmte Reihe (von Variablen): Es sei

$$\eta = x_0 y_1 y_2 y_3 \cdots$$

$$\zeta = x_0 z_1 z_2 z_3 \cdots$$

Ersetzt man dann in $z_1, z_2, \cdots$ die $x_1, x_2, x_3, \cdots$ bzw. durch $y_1, y_2, y_3, \cdots$, so ist das Resultat $(\eta \cdot \zeta)$.

SATZ 1. *Das Produkt von zwei normalen Folgen ist eine normale Folge.*

Beweis: η und ζ werden wie in der Festsetzung 4 bezeichnet und $(\eta \cdot \zeta)$ werde durch $x_0 u_1 u_2 u_3 \cdots$ bezeichnet.

Nach der Definition einer Normalfolge gibt es ein m und ein n sodass 1) $(x_0 y_1 y_2 \cdots y_n)$ eine normale Kombination von $x_1, x_2, \cdots, x_m$ ist, die x_m wirklich enthält, 2) $y_{n+j} \equiv x_{m+j}$. In derselben Weise gibt es ein p und ein q, sodass 1) $(x_0 z_1 z_2 \cdots x_q)$ eine normale Kombination von $x_0 x_1 x_2 \cdots x_p$ ist, die weiterhin x_p wirklich enthält, und 2) $z_{q+j} \equiv x_{p+j}$. Wir können weiter annehmen, dass $p = n$ ist; denn ist $p > n$, so bleibt alles rechtig, das ich über m, n gesagt habe, wenn ich m durch $m + p - n$ ersetze, und ist $p < n$, so kann ich in ähnlicher Weise p durch n, auch q durch $q + n - p$ ersetzen.

Nach diesen Erklärungen sieht man sofort, dass u_i für $i \leq q$ eine Kombination von $x_1, x_2, \cdots, x_m$ ist, während $u_{q+j} \equiv y_{n+j} \equiv x_{m+j}$. Daher ist $x_0 u_1 u_2 \cdots u_{q+1}$ eine normale Kombination von $x_1, x_2, \cdots, x_{m+1}$, und $x_0 u_1 u_2 u_3 \cdots$ ist die durch diese normale Kombination bestimmte Folge, w.z.b.w.

SATZ 2. *Sind Y und Z Kombinatoren, die den normalen Folgen η bzw. ζ von Variablen entsprechen, so entspricht $(Y \cdot Z)$ der Folge $(\eta \cdot \zeta)$.*

Beweis: Sind η und ζ, wie in der Festsetzung 4 bezeichnet, so gibt es m, n, p, q, sodass

$Y x_0 x_1 x_2 \cdots x_m \stackrel{=}{=} x_0 y_1 y_2 \cdots y_n$ x_m nicht ausgelassen
$Z x_0 x_1 x_2 \cdots x_p \stackrel{=}{=} x_0 z_1 z_2 \cdots z_q$ x_p nicht ausgelassen.

Wir können ohne Beschränkung der Allgemeinheit annehmen, dass $n = p$ gilt; denn ist $p > n$, so können wir $x_{m+1}, x_{m+2}, \cdots, x_{m+p-n}$ zu den beiden Seiten der

Convention 3. By a *normal sequence* (of variables) we understand a sequence which is determined by a normal combination of $x_0, x_1, x_2, \cdots, x_n$ (§1, Convention 4), where n is some whole number > 0. Such normal sequences will be denoted with greek letters.

Convention 4. By the *product* $(\eta \cdot \xi)$ of two normal sequences η and ξ we understand the following particular sequences (of variables): Assume

$$\eta = x_0 y_1 y_2 y_3 \ldots$$

$$\xi = x_0 z_1 z_2 z_3 \ldots$$

If in $z_1, z_2, \ldots$ one replaces the $x_1, x_2, x_3 \ldots$ by $y_1, y_2, y_3, \ldots$ respectively, then the result is $(\eta \cdot \xi)$.

Theorem 1. *The product of two normal sequences is a normal sequence.*

Proof. Let η and ξ be denoted as in Convention 4 and denote $(\eta \cdot \xi)$ by
$$x_0 u_1 u_2 u_3 \ldots$$
By the definition of normal sequences there are an m and an n such that 1) $(x_0 y_1 y_2 \ldots y_n)$ is a normal combination of $x_1, x_2, \ldots, x_m$ which really contains x_m, 2) $y_{n+j} \equiv x_{m+j}$. In the same way there are a p and a q such that 1) $(x_0 z_1 z_2 \ldots z_q)$ is a normal combination of $x_0, x_1, x_2, \ldots, x_p$ which furthermore really contains x_p, 2) $z_{q+j} \equiv x_{p+j}$. We can further assume that $p = n$; because if $p > n$ then everything that I said about m and n remains true if I replace m by $m + p - n$, and if $p < n$ then in the same way I can replace p by n and also q by $q + n - p$.

By this explanation one sees that u_i for $i \leqq q$ is a combination of $x_1, x_2, \ldots, x_m$ whereas $u_{q+j} \equiv y_{n+j} \equiv x_{m+j}$. Hence $x_0 u_1 u_2 \ldots u_{q+1}$ is a normal combination of $x_1, x_2, \ldots, x_{m+1}$, and $x_0 u_1 u_2 u_3 \ldots$ is the sequence determined by this normal combination, qed.

Theorem 2. *If Y and Z are combinators which conform to the normal sequences of variables η resp. ξ, then $(Y \cdot Z)$ conforms to the sequence $(\eta \cdot \xi)$.*

Proof. Let η and ξ be denoted as in convention 4; hence there are m, n, p, q, such that

$$Y x_0 x_1 x_2 \cdots x_m \stackrel{=}{=} x_0 y_1 y_2 \cdots y_n \quad x_m \text{ is not omitted}$$
$$Z x_0 x_1 x_2 \cdots x_p \stackrel{=}{=} x_0 z_1 z_2 \cdots z_q \quad x_p \text{ is not omitted}$$

Without loss of generality we can assume that $n = p$; because if $p > n$, we can add $x_{m+1}, x_{m+2}, \cdots, x_{m+p-n}$ to both sides of the first

ersten Gleichung hinzufügen, und ist $p < n$, so können wir x_{p+1}, x_{p+2}, $\cdots$, x_n zu den beiden Seiten der zweiten Gleichung hinzufügen. Dann gilt

$$\begin{aligned}(Y \cdot Z)x_0 x_1 x_2 \cdots x_m &\overset{\sim}{=} Y(Z x_0) x_1 x_2 \cdots x_m \quad \text{(II B 4 Satz 1)} \\ &\overset{\sim}{=} Z x_0 y_1 y_2 y_3 \cdots y_n \\ &\overset{\sim}{=} x_0 u_1 u_2 \cdots u_q,\end{aligned}$$

wo $u_i \equiv z_i$ mit x_i durch y_i ersetzt gilt.

§3. *Die Gruppierungen.*

Festsetzung 1. Eine Folge lauter Variablen heisst eine Gruppierung, wenn die Variablen darin in ihrer ursprünglichen Reihenfolge ohne Wiederholungen oder Auslassungen, aber natürlich in beliebiger Weise in Klammern zusammengefasst, erscheinen. Z. B. sind

$$x_0(x_1 x_2)(x_3(x_4(x_5 x_6)x_7))x_8 x_9 \cdots$$

$$x_0(x_1(x_2(x_3 x_4)x_5 x_6 x_7)x_8)x_9 x_{10}$$

Gruppierungen. Jede Gruppierung ist eine normale Folge.

Festsetzung 2. Unter die Gruppierungen ist die Folge

$$x_0 x_1 x_2 x_3 \cdots$$

einzuschliessen. Diese Gruppierung soll die *identische Gruppierung* heissen. Ihr entspricht der Identitätskombinator I.

Ich werde nun beweisen, dass jeder Gruppierung ein gewisser eindeutig bestimmter Kombinator entspricht.

SATZ 1. *Der Kombinator* $B_m B_n$ ($m \geq 0$, $n > 0$) *entspricht der Gruppierung, welche dann entsteht, wenn man* $x_{m+1}, x_{m+2}, \cdots, x_{m+n+1}$ *in einem einzigen Klammerpaar zusammenfasst. D. h.:*

$$B_m B_n x_0 x_1 x_2 \cdots x_{m+n+1} \overset{\sim}{=} x_0 x_1 x_2 \cdots x_m (x_{m+1} x_{m+2} \cdots x_{m+n+1}).$$

Beweis:
$$B_m B_n x_0 x_1 x_2 \cdots x_{m+n+1} \overset{\sim}{=} B_n(x_0 x_1 x_2 \cdots x_m) x_{m+1} \cdots x_{m+n+1}$$
$$\text{(vgl. II B 1, Satz 3),}$$
$$\overset{\sim}{=} x_0 x_1 x_2 \cdots x_m(x_{m+1} x_{m+2} \cdots x_{m+n+1}) \quad \text{(vgl. II B 1, Satz 3),}$$

SATZ 2. *Jeder Kombinator der Form*
(1) $(B_{m_q} B_{n_q}) \cdot (B_{m_{q-1}} B_{n_{q-1}}) \cdots (B_{m_{q-2}} B_{n_{q-2}}) \cdots (B_{m_2} B_{n_2}) \cdot (B_{m_1} B_{n_1})$
entspricht einer Gruppierung.

Beweis: Folgt aus Satz 1 und §2 Satz 2, weil das Produkt (im Sinne von §2) zweier Gruppierungen wieder eine Gruppierung ist.

equation, and if $p < n$, we can add $x_{p+1}, x_{p+2}, \cdots, x_n$ to both sides of the second equation. Then we have that

$$(Y \cdot Z)x_0 x_1 x_2 \cdots x_m \stackrel{\sim}{=} Y(Zx_0)x_1 x_2 \cdots x_m \quad \text{(II B 4 Theorem 1)}$$
$$\stackrel{\sim}{=} Z x_0 y_1 y_2 \cdots y_n$$
$$\stackrel{\sim}{=} x_0 u_1 u_2 \cdots u_q,$$

where $u_i \equiv z_i$ in which x_i is replaced by y_i.

§3. The groupings

Convention 1. A sequence of variables only is called a *grouping*, if the variables appear in their original numerical order without repetitions or omissions, but naturally enclosed between parenthesis in an arbitrary fashion. E.g. the following

$$x_0(x_1 x_2)(x_3(x_4(x_5 x_6)x_7))x_8 x_9 \cdots$$
$$x_0(x_1(x_2(x_3 x_4)x_5 x_6 x_7)x_8)x_9 x_{10}$$

are groupings. Every grouping is a normal sequence.

Convention 2. The sequence
$$x_0 x_1 x_2 x_3 \cdots$$
is included in the groupings. This grouping shall be called the *identity grouping*. The identity combinator I conforms to it.

I will now show, that every grouping conforms to a uniquely determined combinator.

Theorem 1. *The combinator $B_m B_n$ ($m \geq 0, n > 0$) conforms to the grouping which results from enclosing $x_{m+1}, x_{m+2}, \cdots, x_{m+n+1}$ between one pair of parenthesis. I.e.:*

$$B_m B_n x_0 x_1 x_2 \cdots x_{m+n+1} \stackrel{\sim}{=} x_0 x_1 x_2 \cdots x_m (x_{m+1} x_{m+2} \cdots x_{m+n+1}).$$

Proof.

$$B_m B_n x_0 x_1 x_2 \cdots x_{m+n+1}$$
$$\stackrel{\sim}{=} B_n(x_0 x_1 x_2 \cdots x_m) x_{m+1} x_{m+2} \cdots x_{m+n+1} \quad \text{(cf. II B 1, Thm. 3)}$$
$$\stackrel{\sim}{=} x_0 x_1 x_2 \cdots x_m (x_{m+1} x_{m+2} \cdots x_{m+n+1}) \quad \text{(cf. II B 1, Thm. 3)}$$

Theorem 2. *Every combinator of the form*

(1) $(B_{m_q} B_{n_q}) \cdot (B_{m_{q-1}} B_{n_{q-1}}) \cdot (B_{m_{q-2}} B_{n_{q-2}}) \cdots (B_{m_2} B_{n_2}) \cdot (B_{m_1} B_{n_1})$

conforms to a grouping.

Proof. Follows from Theorem 1 and §2 Theorem 2, because the product (in the sense of §2) of two groupings is again a grouping.

SATZ 3. *Jeder Gruppierung, die nicht die identische ist, entspricht ein und nur ein Kombinator der Form (1) mit*

(2) $$m_q > m_{q-1} > m_{q-2} > \cdots m_2 > m_1.$$

Beweis: Wir nehmen an, dass eine Gruppierung gegeben ist, worin alle die nach I C, Def. 1 fortgeschafften Klammern, sowie auch die die gesamte Gruppierung einschliessenden, wirklich fortgeschafft sind. Die übrig bleibenden Klammern befinden sich in Paaren–eine Anfangsklammer und eine ihr zugehörige Schlussklammer–ein solches Paar nennen wir ein Klammerpaar. Wir bezeichnen dann die Gruppierung mit Γ_q, wo q die Anzahl dieser übrig bleibenden Klammerpaare ist. Es gilt $q \geq 1$, wenn die Gruppierung nicht die identische ist.

Nun sei das Klammerpaar, dessen Anfangsklammer am weitesten links steht, als das erste angesehen. Mit diesem verknüpfen wir die Zahlen m_1, n_1 wie folgt: x_{m_1} soll das letzte x sein, das vor der Anfangsklammer steht, und $n_1 + 1$ soll die Anzahl der innerhalb des Klammerpaares stehenden Glieder sein–wo ein eingeklammerter Teilausdruck, der selbst innerhalb eines anderen Klammerpaares steht, ist als ein einziges Glied des letzteren anzusehen.

Zunächst schaflen wir das erste Klammerpaar aus Γ_q fort. Die so gestaltete Gruppierung nennen wir Γ_{q-1}. Wir suchen dann das erste Klammerpaar in Γ_{q-1}, und bestimmen davon die Zahlen m_2 und n_2 genau so wie die vorigen m_1 und n_1 aus Γ_q bestimmt wurden. Dann schaffen wir dieses Klammerpaar weg und gestalten eine neue Gruppierung Γ_{q-2}, wovon wir die Zahlen m_3 und n_3 bestimmen, u.s.w.

Nachdem wir diesen Prozess q mal wiederholt haben, kommen wir auf einer Γ_0, welche keine Klammern enthält. Dann zeige ich, dass die so konstruierten Zahlen $m_1, m_2, \cdots m_q, n_1, n_2, \cdots n_q$ die Bedingungen des Satzes erfüllen.

Zunächst ist $m_{i+1} > m_i$. Nach der Definition ist $m_{i+1} \geq m_i$, und die Gleichheit ist unmöglich, weil wir alle die nach I C Def. 1 erlaubten Klammerauslassungen ausgeführt haben, und also zwei Anfangklammern an derselben Stelle nicht stehen können.

Zweitens: der Kombinator (1) mit diesem m_i und n_i entspricht dem Γ_q. In der Tat sei γ_r die Gruppierung, der $B_{m_r} B_{n_r}$ nach Satz 1 entspricht, dann folgt aus der Definition der Γ_i, dass

$$\Gamma_{r+1} = \Gamma_r \cdot \gamma_{q-r} \qquad (r = 1, 2, \cdots, q-1)$$
$$\Gamma_1 = \gamma_q$$

gelten. Daher gilt (das Produkt von Folgen ist assoziativ)

Theorem 3. *Every grouping which is not the identity conforms to one and only one combinator of the form (1) where*

(2) $\quad m_q > m_{q-1} > m_{q-2} > \cdots m_2 > m_1.$

Proof. We assume that a grouping is given in which all the parentheses removed according to I C, Def. 1, as well as those which enclose the entire grouping, are really removed. The remaining parentheses occur in pairs–an opening parenthesis and its associated closing parenthesis–we call such a pair a parenthesis-pair. We then call the grouping Γ_q, where q is the number of these remaining parenthesis-pairs. Then $q \geqq 1$, because the grouping is not the identity

Now let the parenthesis-pair whose opening parenthesis occurs furthest to the left, be considered first. We associate to this the numbers m_1, n_1 as follows: x_{m_1} is the last x which occurs before the opening parenthesis and $n_1 + 1$ is the number of elements inside the parenthesis-pair–where a subexpression enclosed in parentheses which is itself inside another parenthesis-pair, is to be seen as a single element of the latter.

First we remove the first parenthesis-pair from Γ_q. We call this remodeled grouping Γ_{q-1}. We then look for the first parenthesis-pair in Γ_{q-1} and determine from it the numbers m_2 and n_2 exactly as the numbers m_1 and n_1 were determined by Γ_q. Then we remove this parenthesis-pair and configure a new grouping Γ_{q-2}, from which we determine the numbers m_3 and n_3, etc.

After we have repeated this process q times, we reach a Γ_0 which contains no parentheses. Then I show that the numbers $m_1, m_2, \cdots m_q$, $n_1, n_2, \cdots n_q$ thus constructed, satisfy the conditions of the theorem.

First $m_{i+1} > m_i$. By definition $m_{i+1} \geqq m_i$, and equality is impossible, because we have performed all the allowable omissions of parenthesis according to I C Def. 1, and therefore two opening parentheses cannot occur in the same position.

Second: the combinator (1) with these m_i and n_i conforms to Γ_q. In fact, let γ_r be the grouping which by Theorem 1 conforms to $B_m B_n$, then by the definition of Γ_i it follows that

$$\begin{aligned} \Gamma_{r+1} &= \Gamma_r \cdot \gamma_{q-r} \quad (r = 1, 2, \cdots, q-1) \\ \Gamma_1 &= \gamma_q \end{aligned}$$

Hence we get (the product of sequences is associative)

$$\Gamma_q = \gamma_q \cdot \gamma_{q-1} \cdots \gamma_1.$$

Daraus folgt die Behauptung nach §2, Satz 2.

Zuletzt gibt es nur einen Kombinator, der die Bedingungen erfüllt. Denn jeder andere Kombinator der Form (1), wofür (2) gilt, entspricht nach dem eben durchgeführten Beweis einer Gruppierung von ganz anderer Klammerstruktur. Aber derselbe Kombinator kann nicht zwei so verschiedenen Folgen entsprechen. (cf. §1, Hilfsatz 3).

§4. *Die Umwandlungen.*

Festsetzung 1. Eine normale Folge von $x_0, x_1, x_2, \cdots, x_n$, worin nach den Auslassungen von I C Def. 1, keine Klammern (ausser den die gesamte Folge einschliessenden) erscheinen, nenne ich eine *Umwandlung*. (Diese Festsetzung stimmt mit der Erklärung im Abschnitte *A* überein). Z. B. sind

$$x_0 x_1 x_3 x_1 x_2 x_4 x_5 \cdots$$

$$x_0 x_2 x_4 x_2 x_3 x_5 x_6 \cdots$$

Umwandlungen, die erste ohne, die zweite mit Auslassungen.

Festsetzung 2. Die Folge:

$$x_0 x_1 x_2 \cdots$$

der der Kombinator I entspricht, ist sowohl eine Umwandlung als auch eine Gruppierung. Ich nenne sie die *Identische Umwandlung*. Um weitere Umschreibungen zu vermeiden, soll hier festgestellt werden, dass diese identische Umwandlung zu allen den hierunter betrachteten Gattungen von Umwandlungen gehört.

SATZ 1. *Jede Umwandlung lässt sich in eindeutiger Weise als Produkt einer Umwandlung κ, die nur Auslassungen zulässt, wie etwa*

(1) $\quad (x_0 x_1 x_2 \cdots x_{h_1-1} x_{h_1+1} x_{h_1+2} \cdots x_{h_2-1} x_{h_2+1} \cdots x_{h_3-1} x_{h_3+1} \cdots x_{h_p-1} x_{h_p+1} \cdots),$
und einer Umwandlung μ ohne Auslassungen darstellen.

Beweis: ω sei die gegebene Umwandlung. Wenn in ω keine Variablen ausgelassen werden, dann gilt $\omega = (\kappa \cdot \mu)$, wo κ die identische Umwandlung ist und $\mu = \omega$ ist. Sonst seien $x_{h_1}, x_{h_2}, \cdots, x_{h_p}$ die aus ω ausgelassenen Variablen. κ sei die Umwandlung (1), mit $h_1 \cdots h_p$ wie eben definiert. μ sei die Umwandlung, welche entsteht, wenn man in ω x_i durch x_j ersetzt, wo j aus i folgendermassen bestimmt wird: wenn $i < h_1$ ist, dann ist $j = i$; wenn $h_k < i < h_{k+1}$ ist, dann ist $j = i - k$; wenn $h_p < i$ ist, dann ist $j = i - p$. Dann ist μ eine Umwandlung ohne Auslassungen und $\omega = (\kappa \cdot \mu)$.

$$\Gamma_q = \gamma_q \cdot \gamma_{q-1} \cdots \cdots \gamma_1.$$

Hence the assertion holds by §2, Theorem 2.

Finally, there is only one combinator which satisfies the conditions. Because every other combinator of the form (1), for which (2) holds, conforms by the proof described above to a grouping of an entirely different parentheses structure. But the same combinator cannot conform to two different sequences. (cf. §1, lemma 3).

§4. *The transformations*

Convention 1. A normal sequence of $x_0, x_1, x_2, \cdots, x_n$, in which, after making the omissions of parenthesis of I C Def. 1, no parentheses occur (except those which enclose the entire sequence), I call a *Transformation (Umwandlung)*. (This convention coincides with the explanation in section A). E.g.

$$x_0 x_1 x_3 x_1 x_2 x_4 x_5 \cdots$$

$$x_0 x_2 x_4 x_2 x_3 x_5 x_6 \cdots$$

are transformations, the first without, the second with omissions.

Convention 2. The sequence $x_0 x_1 x_2 \cdots$ which conforms to the combinator I, is both a transformation and also a grouping. I call it the *identity transformation*. In order to avoid additional descriptions, it will be established here that this identity transformation belongs to all the genera of transformations considered below.

Theorem 1. *Every transformation can be represented in a unique way as a product of a transformation κ which only admits omissions like this*

(1) $(x_0 x_1 x_2 \cdots x_{h_1-1} x_{h_1+1} x_{h_1+2} \cdots x_{h_2-1} x_{h_2+1} \cdots x_{h_3-1}$
$ x_{h_3+1} \cdots x_{h_p-1} x_{h_p+1} \cdots),$

and a transformation μ without omissions.

Proof. Let ω be the given transformation. If there are no omitted variables in ω, then $\omega = (\kappa \cdot \mu)$, where κ is the identity transformation and $\mu = \omega$. Otherwise let $x_{h_1}, x_{h_2}, \cdots, x_{h_p}$ be the variables omitted from ω. Let κ be the transformation (1), with $h_1 \cdots h_p$ defined as above. Let μ be the transformation which results when one replaces x_i in ω by x_j where j is determined from i in the following way: If $i < h_1$ then $j = i$; if $h_k < i < h_{k+1}$, then $j = i - k$; if $h_p < i$, then $j = i - p$. Hence μ is a transformation without omissions and $\omega = (\kappa \cdot \mu)$.

κ' sei nun irgendeine Umwandlung der Form (1) (bzw. die identische Umwandlung) und μ' sei eine Umwandlung ohne Auslassungen. Es sei $\omega' = (\kappa' \cdot \mu')$. Bilden wir κ'' und μ'' aus ω' genau wie wir κ und μ aus ω gebildet haben, so ist $\kappa'' = \kappa'$ und $\mu'' = \mu'$. Also wenn $\omega' = \omega$ gilt, so ist $\kappa' = \kappa$ und $\mu' = \mu$. Also sind κ und μ durch ω eindeutig bestimmt.

SATZ 2. *Jedem κ, das nicht das identische ist, entspricht ein und nur ein Kombinator der Form*

(2) $$K_{h_p} \cdot K_{h_{p-1}} \cdots K_{h_2} \cdot K_{h_1},$$

wo

(3) $$h_1 < h_2 < \cdots < h_{p-1} < h_p$$

gilt.

Beweis: Es sei eine Umwandlung κ der Form (1) gegeben. Der Kombinator (2) mit dem durch (1) bestimmten $h_1, h_2, \cdots, h_p$ entspricht dann diesem κ, und die Bedingung (3) ist natürlich erfüllt. Irgendein anderer Kombinator (2), wofür (3) erfüllt ist, entspricht nach dem eben Gesagten einer von κ verschiedenen Folge κ', also nicht zu κ (§1, Hilfssatz 3).

Festsetzung 3. Unter einer *Permutationsfolge* verstehen wir eine normale Folge, die durch eine Permutation bestimmt ist, oder, was dasselbe ist, eine Umwandlung ohne Auslassungen oder Wiederholungen.

SATZ 3. *Jede Umwandlung ohne Auslassungen lässt sich als Produkt zweier Faktoren darstellen, wovon der zweite eine Permutationsfolge ist, während im ersten die Variablen ihre ursprüngliche Reihenfolge behalten, aber wiederholt werden können. Dieser erste Faktor ist eindeutig bestimmt.*

Beweis: μ sei eine gegebene Umwandlung ohne Auslassungen. Wenn es in μ keine wiederholten Variablen gibt, so ist der zweite Faktor μ selbst, der erste die identische Umwandlung. Sonst seien $x_{k_1}, x_{k_2}, \cdots, x_{k_q}$ ($k_1 < k_2 < \cdots < k_q$) sämtliche in μ wiederholte Variablen, und wir setzen fest, dass x_{k_1} $(r_1 + 1)$ mal, x_{k_2} $(r_2 + 1)$ mal u. s. w. bis x_{k_q} $(r_q + 1)$ mal in μ erscheinen. Dann betrachten wir die Umwandlung:

(4) $(x_0 x_1 x_2 \cdots x_{k_1-1} x_{k_1} x_{k_1} \cdots (r_1 + 1)$ mal $\cdots x_{k_1} x_{k_1+1} \cdots x_{k_2-1} x_{k_2} x_{k_2} \cdots$
$(r_2 + 1)$ mal $\cdots x_{k_2} x_{k_2+1} \cdots x_{k_q-1} x_{k_q} x_{k_q} \cdots (r_q + 1)$ mal
$\cdots x_{k_q} x_{k_q+1} \cdots)$.

Dann ist μ durch eine Permutation der in (4) erscheinenden Zeichen bestimmt, also ist es das Produkt von (4) und der durch diese Permutation bestimmten Permutationsfolge.

Umgekehrt sei eine Kombination (4) gegeben (wo wir unter $q = 0$ die

Assume now κ' is an arbitrary transformation of the form (1) (resp. the identity transformation) and μ' a transformation without omissions. Let $\omega' = (\kappa' \cdot \mu')$. We build κ'' and μ'' from ω' exactly as we have built κ and μ from ω, hence $\kappa'' = \kappa'$ and $\mu'' = \mu'$. Therefore when $\omega' = \omega$, then $\kappa' = \kappa$ and $\mu' = \mu$. Hence κ and μ are uniquely determined by ω.

Theorem 2. *Every such κ which is not the identity conforms to one and only one combinator of the form*

(2) $\qquad K_{h_p} \cdot K_{h_{p-1}} \cdots K_{h_2} \cdot K_{h_1},$

where

(3) $\qquad h_1 < h_2 < \cdots < h_{p-1} < h_p.$

Proof. Assume κ is a given transformation of the form (1). The combinator (2) with the $h_1, h_2, \cdots, h_p$ determined by (1) conform then to this κ, and the condition (3) is naturally satisfied. Any other combinator (2), for which (3) is satisfied, conforms by what is said above to a sequence κ' different from κ, hence not to κ (§1, Lemma 3).

Convention 3. By a *permutation sequence* we understand a normal sequence which is determined by a permutation, or which is the same, a transformation without omissions or repetitions.

Theorem 3. *Every transformation without omissions can be represented as a product of two factors, the second of which is a permutation sequence, whereas in the first the variables are kept in their original order but can be repeated. This first factor is uniquely determined.*

Proof. Let μ be a given transformation without omissions. If no variables are repeated in μ then the second factor is μ itself and the first is the identity transformation. Otherwise, let $x_{k_1}, x_{k_2}, \cdots, x_{k_q}$ ($k_1 < k_2 < \cdots < k_q$) be all the repeated variables in μ, and assume that x_{k_1}, x_{k_2}, etc. to x_{k_q} appear $(r_1 + 1)$, $(r_2 + 1)$, etc. $(r_q + 1)$ times in μ. Then we consider the transformation:

(4) $\quad (x_0 x_1 x_2 \cdots x_{k_1 - 1} \overbrace{x_{k_1} \cdots x_{k_1}}^{(r_1+1) \text{ times}} x_{k_1+1} \cdots x_{k_2 - 1} \overbrace{x_{k_2} \cdots x_{k_2}}^{(r_2+1) \text{ times}} x_{k_2+1}$
$\cdots x_{k_q - 1} \overbrace{x_{k_q} \cdots x_{k_q}}^{(r_q+1) \text{ times}} x_{k_q+1} \cdots)$

Hence μ is determined by a permutation of the symbols appearing in (4), thus it is the product of (4) and the permutation sequence determined by this permutation.

Conversely, assume a given combination (4) (where when $q = 0$ we understand the identity

identische Umwandlung zu verstehen haben). Denn das Produkt von (4) nach irgendeiner Permutationsfolge ist ein μ, worin x_{k_i} ($i = 1, 2, \cdots, q$) ($r_i + 1$) mal erscheint und kein anderes x wiederholt ist. Also kann ein μ nie zugleich als Produkt von zwei verschiedenen Kombinationen der Form (4) mit Permutationsfolgen dargestellt werden.

Der Satz wird nun bewiesen, wenn wir bemerken, dass wenn $q, k_1, k_2, \cdots k_q$ beliebig sind, (4) die allgemeinste, den Bedingungen für den ersten Faktor genügende Folge ist.

Def. 1.
$$W_k^1 \equiv W_k \qquad k = 1, 2, 3, \cdots,$$
$$W_k^{r+1} \equiv W_k \cdot W_k^r \quad k = 1, 2, 3, \cdots, \quad r = 1, 2, 3, 4, \cdots.$$

SATZ 4. *Es gibt einen und nur einen Kombinator der Form*
$$(5) \qquad W_{k_q}^{r_q} \cdot W_{k_{q-1}}^{r_{q-1}} \cdots W_{k_2}^{r_2} \cdot W_{k_1}^{r_1},$$
wo ferner
$$(6) \qquad k_1 < k_2 < \cdots < k_q$$

gilt, der einer gegebenen von der identischen Umwandlung verschiedenen, den Bedingungen für den ersten Faktor im Satze 3 genügenden Folge entspricht.*

Beweis: Zuerst: W_k^r entspricht der Folge
$$(x_0 x_1 x_2 \cdots x_{k-1} x_k x_k x_k \cdots (r+1) \text{ mal} \cdots x_k x_{k+1} x_{k+2} \cdots).$$

In der Tat für $r = 1$ folgt dies aus II B 3, Satz 4. Ist es für ein gegebenes r angenommen, dann haben wir für $r+1$
$$W_k^{r+1} x_0 x_1 x_2 \cdots x_k \ \tilde{=}\ W_k^r x_0 x_1 \cdots x_{k-1} x_k x_k$$
$$\tilde{=}\ x_0 x_1 x_2 \cdots x_{k-1} x_k x_k \cdots (r+2) \text{mal} \cdots x_k.$$

Es wird nun bewiesen werden, dass, wenn (6) gilt, (5) wie es geschrieben steht dem Ausdruck (4) entspricht. Zu diesem Behuf kürzen wir (5) mit q durch s ersetzt mit $\mathfrak{W}_s$, und den Ausdruck
$$(x_0 x_1 \cdots x_{k_1-1} x_{k_1} x_{k_1} \cdots (r_1+1) \text{mal}$$
$$\cdots x_{k_1} x_{k_1+1} \cdots x_{k_s-1} x_{k_s} x_{k_s} \cdots (r_s+1) \text{mal} \cdots x_{k_s})$$

mit X_s ab. Dann haben wir schon für $s = 1$ bewiesen,
$$(7) \qquad \mathfrak{W}_s x_0 x_1 x_2 \cdots x_{k_s} \tilde{=} X_s$$

Ist dies für ein bestimmtes s vorausgesetzt, so haben wir

---

*Dieser Satz und Lemma 3 meiner oben zit. Abhandlung sind wesentlich äquivalent. Der hier gegebene Beweis ist alternativ zu jenem.

transformation). Then the product of (4) after an arbitrary permutation sequence is a μ, in which x_{k_i} ($i = 1, 2, \cdots, q$) appears $(r_i + 1)$ times and no other x is repeated. Hence a μ can never be represented at once as a product of two different combinations of the form (4) with permutation sequences.

The theorem will now be proven when we realize that if $q, k_1, k_2, \cdots, k_q$ are arbitrary, then (4) is the most general sequence which satisfies the conditions for the first factor.

Def. 1.
$$W_k^1 \equiv W_k \qquad k = 1, 2, 3, \cdots,$$
$$W_k^{r+1} \equiv W_k \cdot W_k^r \quad k = 1, 2, 3, \cdots, \quad r = 1, 2, 3, 4, \cdots.$$

Theorem 4. *There is one and only one combinator of the form*

(5) $\qquad W_{k_q}^{r_q} \cdot W_{k_{q-1}}^{r_{q-1}} \cdots W_{k_2}^{r_2} \cdot W_{k_1}^{r_1},$

where further

(6) $\qquad k_1 < k_2 < \cdots < k_q,$

this given combinator that comforms to a given sequence, different from the identity transformation, that satisfies the conditions of the first factor in Theorem 3 [as a sequence].*

Proof. First: W_k^r conforms to the sequence

$$(x_0 x_1 x_2 \cdots x_{k-1} \overbrace{x_k \cdots x_k}^{(r+1) \text{ times}} x_{k+1} x_{k+2} \cdots).$$

In fact for $r = 1$ this follows from II B 3, Theorem 4. Assume it holds for some r, then we have for $r+1$

$$W_k^{r+1} x_0 x_1 x_2 \cdots x_k \;\dot{=}\; W_k^r x_0 x_1 \cdots x_{k-1} x_k x_k$$
$$\dot{=}\; x_0 x_1 x_2 \cdots x_{k-1} \overbrace{x_k \cdots x_k}^{(r+2) \text{ times}}.$$

It will now be proven that when (6) holds, (5) as written conforms to expression (4). For this purpose, we abbreviate (5) where q is replaced by s as $\mathfrak{W}_s$, and the expression

$$(x_0 x_1 \cdots x_{k_1-1} \overbrace{x_{k_1} \cdots x_{k_1}}^{(r_1+1) \text{ times}} x_{k_1+1} \cdots x_{k_s-1} \overbrace{x_{k_s} \cdots x_{k_s}}^{(r_s+1) \text{ times}})$$

as X_s. Then we have already shown for $s = 1$ that,

(7) $\qquad \mathfrak{W}_s x_0 x_1 x_2 \cdots x_{k_s} \;\dot{=}\; X_s$

If we assume this for some s, then we have

---
*This theorem and Lemma 3 of my above cited paper are essentially equivalent. The proof given here is an alternative to that one.

$$\mathfrak{W}_{s+1}x_0x_1x_2\cdots x_{k_s+1} \overset{\sim}{=} W^{r_s+1}_{k_s+1}(\mathfrak{W}_s x_0)x_1x_2\cdots x_{k_s+1}$$
$$\overset{\sim}{=} \mathfrak{W}_s x_0 x_1 x_2 \cdots x_{k_{s+1}-1}x_{k_{s+1}}x_{k_s+1}\cdots (r_{s+1})\text{mal}\cdots x_{k_s+1}$$
(nach dem eben bewiesenen),

$$\overset{\sim}{=} X_s x_{k_s+1}x_{k_s+2}\cdots x_{k_{s+1}-1}x_{k_{s+1}}x_{k_s+1}\cdots (r_{s+1}+1)\text{mal}\cdots x_{k_s+1}$$
(nach der Voraussetzung),

$$\overset{\sim}{=} X_{s+1} \qquad \text{(nach der Festsetzung über } X_s\text{).}$$

Also wird durch Induktion (7) für $s = q$, also die Behauptung bewiesen.

Der Beweis des Satzes folgt gleich. Denn wenn wir die Konstanten $q, k_1, k_2, \cdots, k_q$ in (5) einsetzen, so entspricht der resultierende Kombinator der Folge (4) nach dem letzten Absatz. Wenn wir andere Konstanten, die (6) genügen, in (5) einsetzen, so entspricht der resultierende Kombinator einer ganz anderen Folge der Form (4). Also gibt es nur einen Kombinator der betreffenden Beschaffenheit.

SATZ 5. *Jeder Permutationsfolge entspricht ein Kombinator* $\mathfrak{C}$, *der aus einem Produkt lauter* $C_1, C_2 \cdots$ *besteht, und zwar so, dass das mit dem höchsten Index versehene* C_n *nur einmal vorkommt.*

Beweis: Nach einem wohlbekannten Satz über Permutationen ist jede Permutation der Elemente $x_1, x_2, \cdots, x_m$ ein Produkt von Transformationen benachbarter Elementen. Dies bedeutet, in unsere Terminologie übersetzt, dass jede Permutationsfolge, die durch eine, Permutation von $x_1, x_2, \cdots, x_m$ bestimmt wird, ein Produkt der Folgen

$$(x_0 x_2 x_1 x_3 \cdots)$$
$$(x_0 x_1 x_3 x_2 \cdots)$$
$$\cdots\cdots$$
$$(x_0 x_1 x_2 \cdots x_{m-2} x_m x_{m-1} x_{m+1} \cdots)$$

ist. Diesen Folgen entsprechen bzw. die Kombinatoren $C_1, C_2, \cdots, C_{m-1}$. Infolgedessen entspricht der gegebenen Permutationsfolge ein $\mathfrak{C}$, das aus einem Produkt von lauter $C_1, C_2, \cdots, C_{m-1}$ besteht (§2, Satz 2).

Es sei nun eine PeImutationsfolge π gegeben, die x_{m+1}, aber keine mit höherem Index versehene Variable, wirklich permutiert. x_k sei die Variable, die die Stelle von x_{m+1} einnimmt. Dann ist π ein Produkt von zwei Folgen π_1, und π_2, wo

$$\pi_1 = (x_0 x_1 x_2 \cdots x_{k-1} x_{k+1} \cdots x_{m-1} x_m x_{m+1} x_k x_{m+2} \cdots)$$

ist und π_2 durch eine Permutation von $x_1 x_2 \cdots x_m$ bestimmt ist. Der Folge π_1 entspricht aber der Kombinator $\mathfrak{C}_1$,

$$\mathfrak{C}_1 \equiv C_k \cdot C_{k+1} \cdot \cdots \cdot C_{m-1} \cdot C_m.$$

$\mathfrak{W}_{s+1}x_0x_1x_2\cdots x_{k_{s+1}}$
$\doteq W_{k_{s+1}}^{r_{s+1}}(\mathfrak{W}_s x_0)x_1x_2\cdots x_{k_{s+1}}$
$\doteq \mathfrak{W}_s x_0 x_1 x_2 \cdots x_{k_{s+1}-1} \overbrace{x_{k_{s+1}} \cdots x_{k_{s+1}}}^{(r_{s+1})\text{times}}$ (by the proven above),
$\doteq X_s x_{k_s+1} x_{k_s+2} \cdots x_{k_{s+1}-1} \overbrace{x_{k_{s+1}} \cdots x_{k_{s+1}}}^{(r_{s+1}+1)\text{times}}$ (by the hypothesis),
$\doteq X_{s+1}$ (by the definition of X_s).

Hence (7) holds for $s = q$, so by induction the assertion is proven.

The proof of the theorem follows similarly. Because when we place the constants $q, k_1, k_2, \cdots, k_q$ in (5), then by the last paragraph the resulting combinator conforms to sequence (4). If in (5) we place other constants which satisfy (6), then the resulting combinator conforms to an entirely different sequence of the form (4). Thus there is only one combinator with the quality in question.

Theorem 5. *Every permutation sequence conforms to one combinator $\mathfrak{C}$ which consists of a product of only $C_1, C_2, \cdots$ such that the C_n with the highest index n only appears once.*

Proof. By a well known theorem on permutations, every permutation of the elements $x_1, x_2, \cdots, x_m$ is a product of transformations of adjacent elements. This means, translated into our terminology, that every permutation sequence which is determined by the permutation of $x_1, x_2, \cdots, x_m$ is a product of the sequences:

$$(x_0 x_2 x_1 x_3 \cdots)$$
$$(x_0 x_1 x_3 x_2 \cdots)$$
$$\cdots\cdots$$
$$(x_0 x_1 x_2 \cdots x_{m-2} x_m x_{m-1} x_{m+1} \cdots).$$

These sequences conform resp. to the combinators $C_1, C_2, \cdots, C_{m-1}$. Consequently the given permutation sequence conforms to a $\mathfrak{C}$ which is a product of only $C_1, C_2, \cdots, C_{m-1}$ (§2, Theorem 2).

Assume now a given permutation sequence π which really permutes x_{m+1} but no other variable with a higher index. Let x_k be the variable which occupies the place of x_{m+1}. Then π is a product of two sequences π_1 and π_2 where

$$\pi_1 = (x_0 x_1 x_2 \cdots x_{k-1} x_{k+1} \cdots x_{m-1} x_m x_{m+1} x_k x_{m+2} \cdots)$$

and π_2 is determined by a permutation of $x_1 x_2 \cdots x_m$. But the sequence π_1 conforms to the combinator $\mathfrak{C}_1$,

$$\mathfrak{C}_1 \equiv C_k \cdot C_{k+1} \cdot \cdots \cdot C_{m-1} \cdot C_m.$$

Der Folge π_2 entspricht weiter nach dem vorigen Absatz ein $\mathfrak{C}_2$, das ein Produkt lauter $C_1, C_2, \cdots C_{m-1}$ ist. Also entspricht $\mathfrak{C} \equiv \mathfrak{C}_1 \cdot \mathfrak{C}_2$ der Folge π, und $\mathfrak{C}$ erfüllt die Bedingungen des Satzes, weil C_m nur einmal vorkommt.

§5. *Darstellung der allgemeinen normalen Folge.*
SATZ 1. *Jede normale Folge lässt sich in eindeutiger Weise als Produkt einer Umwandlung und einer Gruppierung darstellen.*

Beweis: η sei die gegebene Folge. Wir erzeugen aus η eine Umwandlung ω und eine Gruppierung γ folgendermassen: zuerst schaffen wir alle die innerhalb η erscheinenden Klammern fort, dann soll der resultierende Ausdruck ω heissen. Zweitens lassen wir in η die Klammern stehen und schaffen die Variablen fort, und füllen dann die Leerstellen, wo Variablen früher waren, mit $x_0, x_1, x_2, \cdots$ von links nach rechts in ihrer naturgemässen Reihenfolge, ohne Auslassungen oder Wiederholungen aus. Der neue Ausdruck ist eine Gruppierung, γ. Diese ω und γ nennen wir die mit η assoziierte Umwandlung bzw. Gruppierung. Nach der Festsetzung 4, §2 gilt $\eta = (\omega \cdot \gamma)$.

Nun sei ω' irgendeine Umwandlung und γ' eine Gruppierung. Es sei $\eta' = (\omega' \cdot \gamma')$. Dann sind die mit η' assoziierte Gruppierung bzw. Umwandlung genau dieses ω' bzw. γ'. Infolgedessen muss, wenn $\eta' = \eta$ ist, auch $\omega' = \omega$ und $\gamma' = \gamma$ sein.

SATZ 2. *Jeder normalen Folge entspricht mindestens ein Kombinator der Form*:

$$(\mathfrak{K} \cdot \mathfrak{W} \cdot \mathfrak{C} \cdot \mathfrak{B}),$$

wo a) $\mathfrak{K}$ *in der Form von §4 Satz 2 steht,*
b) $\mathfrak{W}$ *in der Form von §4 Satz 4 steht,*
c) $\mathfrak{C}$ *in der Form von §4 Satz 5 steht,*
d) $\mathfrak{B}$ *in der Form von §3 Satz 3 steht.*
Ferner sind $\mathfrak{K}, \mathfrak{W}$ *und* $\mathfrak{B}$ *durch diese Bedingungen eindeutig bestimmt.*

Beweis: η sei eine gegebene normale Folge. Nach Satz 1 und §4, Sätzen 1, 3 gibt es eine Gruppierung γ, eine Umwandlung κ, die nur Auslassungen zulässt, eine Umwandlung ω, die nur Wiederholungen zulässt, und eine Permutationsfolge, π, derart, dass

$$\eta = (\kappa \cdot \omega \cdot \pi \cdot \gamma)$$

ist. Nach §3 Satz 3, §4 Sätze 2, 4, 5 gibt es gewisse die Bedingungen a-d erfüllende $\mathfrak{K}, \mathfrak{W}, \mathfrak{C}, \mathfrak{B}$ die diesen κ bzw. ω bzw. π bzw. γ entsprechen. Dann entspricht $(\mathfrak{K} \cdot \mathfrak{W} \cdot \mathfrak{C} \cdot \mathfrak{B})$ der Folge η nach §2 Satz 2.

The sequence π_2 conforms further by the above paragraph to a $\mathfrak{C}_2$ which is a product of only $C_1, C_2, \cdots, C_{m-1}$. Thus, $\mathfrak{C} \equiv \mathfrak{C}_1 \cdot \mathfrak{C}_2$ conforms to the sequence π, and $\mathfrak{C}$ satisfies the conditions of the theorem, because C_m only appears once.

§5. Presentation of general normal sequences

Theorem 1. *Every normal sequence can be represented in a unique way as a product of a transformation and a grouping.*

Proof. Let η be the given sequence. We generate from η a transformation ω and a grouping γ as follows: First we remove all the inside parentheses appearing in η, then the resulting expression shall be called ω. Second, we keep the parentheses in η and remove the variables and then fill the blanks where variables appeared earlier with $x_0, x_1, x_2, \cdots$ from left to right in their natural order, without omissions or repetitions. The new expression is a grouping γ. We call this ω and γ respectively the transformation and grouping associated with η. By Convention 4, §2, it holds that $\eta = (\omega \cdot \gamma)$.

Now let ω' be an arbitrary transformation and γ' be a grouping. Let $\eta' = (\omega' \cdot \gamma')$. Then the transformation and grouping associated with η' are respectively exactly this ω' and γ'. Consequently, when $\eta' = \eta$, we must also have $\omega' = \omega$ and $\gamma' = \gamma$.

Theorem 2. *Every normal sequence conforms to at least one combinator of the form:*

$$(\mathfrak{K} \cdot \mathfrak{W} \cdot \mathfrak{C} \cdot \mathfrak{B})$$

where a) $\mathfrak{K}$ *is in the form of §4 Theorem 2,*
 b) $\mathfrak{W}$ *is in the form of §4 Theorem 4,*
 c) $\mathfrak{C}$ *is in the form of §4 Theorem 5,*
 d) $\mathfrak{B}$ *is in the form of §3 Theorem 3.*
Further $\mathfrak{K}, \mathfrak{W}$ and $\mathfrak{B}$ are uniquely determined by these conditions.

Proof. Let η be a given normal sequence. By Theorem 1 and §4, Theorems 1, 3, there exist a grouping γ, a transformation κ which only allows omissions, a transformation ω which only allows repetitions, and a permutation sequence π, such that

$$\eta = (\kappa \cdot \omega \cdot \pi \cdot \gamma)$$

By §3 Theorem 3, §4 Theorems 2, 4, 5 there are $\mathfrak{K}, \mathfrak{W}, \mathfrak{C}$ and $\mathfrak{B}$ which satisfy conditions a–d and which conform to these $\kappa, \omega, \pi, \gamma$ resp. Then, $(\mathfrak{K} \cdot \mathfrak{W} \cdot \mathfrak{C} \cdot \mathfrak{B})$ conforms to η by §2 Theorem 2.

Nun sei irgendeine der Folge η entsprechende Kombinator der Form (1) gegeben, etwa $(\mathfrak{K}' \cdot \mathfrak{W}' \cdot \mathfrak{C}' \cdot \mathfrak{B}')$. Es seien $\kappa', \omega', \pi', \gamma'$, die zu $\mathfrak{K}', \mathfrak{W}', \mathfrak{C}'$ bzw. $\mathfrak{B}'$ gehörenden Folgen; sie gehören denselben Kombinationsgattungen wie κ, ω, π bzw. γ an. Der betreffende Kombinator entspricht dann $(\kappa' \cdot \omega' \cdot \pi' \cdot \gamma')$ (§2, Satz 2); also, wenn er auch dem η entspricht, gilt

$$(\kappa' \cdot \omega' \cdot \pi' \cdot \gamma') = (\kappa \cdot \omega \cdot \pi \cdot \gamma) \qquad \text{(§1 Hilfsatz 3)}.$$

Also gelten $\gamma' = \gamma$ und $(\kappa' \cdot \omega' \cdot \pi) = (\kappa \cdot \omega \cdot \pi')$ (Satz 1); $\kappa' = \kappa$ und $(\omega' \cdot \pi') = (\omega \cdot \pi)$ (§4 Satz 1); $\omega' = \omega$ (§4 Satz 3). Daher sind $\mathfrak{B}'$ und $\mathfrak{B}$ identisch (§3 Satz 3); $\mathfrak{K}'$ und $\mathfrak{K}$ sind identisch (§2 Satz 2); $\mathfrak{W}'$ und $\mathfrak{W}$ sind identisch (§4 Satz 4). Also sind $\mathfrak{K}, \mathfrak{W}$, und $\mathfrak{B}$ durch die Bedingungen eindeutig bestimmt.

D. REGULÄRE KOMBINATOREN.

§1. *Vorläufige Festsetzungen und Sätze.*

Festsetzung 1. Ein Kombinator X heisst *regulär*, wenn er die Form

$$(X_1 \cdot X_2 \cdot \ldots \cdot X_n)$$

hat, wo jedes X_i ferner von einer der Formen

$$B_p B_q, C_q, W_q, K_q, B_p I$$

ist. Die einzelne X_i heissen die *Glieder* von X.

Festsetzung 2. Ein Kombinator heisst *normal*, bzw. *in der normalen Form*, wenn er in der in II C 5 Satz 2 besprochenen Form steht.

Festsetzung 3. Ein Kombinator X heisst in einer gegebenen Form *umformbar*, wenn es ein schon in der betreffenden Form stehendes X' gibt, sodass $\vdash X = X'$.

In diesem Abschnitte beweise ich den Hauptsatz: wenn immer zwei reguläre Kombinatoren X und Y derselben Folge entsprechen, dann $\vdash X = Y$. Dies folgt daraus, dass erstens jeder reguläre Kombinator sich in die Normalform umformen lässt, und zweitens der Hauptsatz gilt, wenn nur X und Y normal sind. Der zweite in Abschnitt A erwähnte Hauptsatz wird hier für normale Kombinationen bewiesen.

Festsetzung 4. Zum Zwecke der Abkürzung möchte ich die folgenden Buchstaben für gewisse Gattungen regulärer Kombinatoren gebrauchen, derart, dass besondere Kombinatoren dadurch bezeichnet werden, dass Indizes an das betreffende Gattungszeichen angeheftet werden.

Now assume a given combinator of the form (1) which conforms to the sequence η, say $(\mathfrak{K}' \cdot \mathfrak{W}' \cdot \mathfrak{C}' \cdot \mathfrak{B}')$. Let κ', ω', π', γ' be the sequences which correspond to $\mathfrak{K}'$, $\mathfrak{W}'$, $\mathfrak{C}'$ resp. $\mathfrak{B}'$; they belong to the same combination genera as κ, ω, π resp. γ. Then the combinator in question conforms to $(\kappa' \cdot \omega' \cdot \pi' \cdot \gamma')$ (§2, Theorem 2); Hence, when it also conforms to η, we have

$$(\kappa' \cdot \omega' \cdot \pi' \cdot \gamma') = (\kappa \cdot \omega \cdot \pi \cdot \gamma) \qquad \text{(§1 Lemma 3)}.$$

Therefore $\gamma' = \gamma$ and $(\kappa' \cdot \omega' \cdot \pi') = (\kappa \cdot \omega \cdot \pi)$ (Theorem 1); $\kappa' = \kappa$ and $(\omega' \cdot \pi') = (\omega \cdot \pi)$ (§4 Theorem 1); $\omega' = \omega$ (§4 Theorem 3). Hence $\mathfrak{B}'$ and $\mathfrak{B}$ are identical (§3 Theorem 3); $\mathfrak{K}'$ and $\mathfrak{K}$ are identical (§2 Theorem 2); $\mathfrak{W}'$ and $\mathfrak{W}$ are identical (§4 Theorem 4). Thus $\mathfrak{K}$, $\mathfrak{W}$ and $\mathfrak{B}$ are uniquely determined by the conditions.

D. Regular Combinators

§1. *Preliminary conventions and theorems.*

Convention 1. A combinator X is called *regular* when it is of the form

$$(X_1 \cdot X_2 \cdot \cdots \cdot X_n)$$

where further, every X_i is of one of the forms

$$B_p B_q, C_q, W_q, K_q, B_p I$$

These individual X_i's are called the *limbs* of X.

Convention 2. A combinator is called *normal*, resp. in *normal form*,[21] if it is in the form discussed in II C 5 Theorem 2.

Convention 3. A combinator X is called *convertible* to a given form if there is an X' already in the form in question such that $\vdash X = X'$.

In this section I will prove the main theorem: whenever two regular combinators X and Y conform to the same sequence then $\vdash X = Y$. This follows from the fact that, first, every regular combinator can be converted into normal form, and secondly, the main theorem holds when X and Y are normal. The second main theorem mentioned in section A will be proven here for normal combinations.

Convention 4. For the sake of brevity I shall use the following letters for certain genera of regular combinators, in such a way that special combinators are denoted thereby by attaching indices to the respective characters.

$\mathfrak{B}$: sämtliche Glieder der Form $B_m B_n$ oder $B_m I$.
$\mathfrak{C}$: sämtliche Glieder der Form C_n oder $B_m I$
$\mathfrak{K}$: sämtliche Glieder der Form K_n oder $B_m I$
$\mathfrak{M}$: sämtliche Glieder der Form W_n, C_n oder $B_m I$
$\mathfrak{W}$: sämtliche Glieder der Form W_n oder $B_m I$
Ω: sämtliche Glieder der Form C_n, K_n, W_n, oder $B_m I$.

SATZ 1. *Zu jedem regulären Kombinator X gibt es ein X' derart, dass 1) X' regulär ist, 2) X' wenn von I selbst verschieden, gar keine Glieder der Form $B_m I$ enthält, während die anderen Glieder genau dieselben wie in X sind, und 3) $\vdash X = X'$.*

Beweis: Klar aus II B2, Satz 1 und B4, Satz 4. Wenn sämtliche Glieder in X der Form $B_m I$ sind, so ist X' gleich I, sonst ist X' von I verschieden.

SATZ 2. $\vdash BB \cdot C_1 = C_1 \cdot C_2 \cdot B$.

Beweis: Wir haben zunächst aus Ax. $(CC)_1$ und II B2, Satz 1

$$\begin{aligned} \vdash C_1 \cdot C_1 &= B_2 I = I \\ \vdash C_2 \cdot C_2 &= B(C_1 \cdot C_1) \quad \text{(II B 4, Satz 2; II B 3, Def. 1)}. \\ &= B_3 I = I. \quad \text{(Ax. } (CC)_1\text{)}. \end{aligned}$$

Daher:

$$\begin{aligned} \vdash BB \cdot C_1 &= C_1 \cdot C_2 \cdot C_2 \cdot C_1 \cdot BB \cdot C_1 \\ &= C_1 \cdot C_2 \cdot B \cdot C_1 \cdot C_1 \quad \text{(Ax. } (BC)\text{)}, \\ &= C_1 \cdot C_2 \cdot B. \quad \text{w.z.b.w.} \end{aligned}$$

§2. *Die kommutativen Gesetze.*

Festsetzung 1. Eine Gleichung der Form

$$\vdash C_1 B_{m+1} X = BX \cdot B_n$$

heisst ein *kommutatives Gesetz* für X, weil es eine gewisse Art von Vertauschbarkeit von X mit anderen Etwasen gewährt. Einige der Axiome sind von dieser Form; diese habe ich kommutative Axiome genannt.

SATZ 1. *Wenn X ein Etwas ist, wofür*

$$\vdash CB_{m+1}X = BX \cdot B_n;$$

dann gilt für irgendein Etwas Y

$$\vdash B_m Y \cdot X = X \cdot B_n Y.$$

Beweis: Aus der Hp. und I D Satz 3 folgt

$\mathfrak{B}$: all its limbs are of the form $B_m B_n$ or $B_m I$.
$\mathfrak{C}$: all its limbs are of the form C_n or $B_m I$.
$\mathfrak{K}$: all its limbs are of the form K_n or $B_m I$.
$\mathfrak{M}$: all its limbs are of the form W_n, C_n or $B_m I$.
$\mathfrak{W}$: all its limbs are of the form W_n or $B_m I$.
Ω: all its limbs are of the form C_n, K_n, W_n or $B_m I$.

Theorem 1. *To every regular combinator X there is an X' such that 1) X' is regular, 2) X', when different from I, has no limb of the form $B_m I$, whereas the other limbs are exactly the same as those in X, and 3) $\vdash X = X'$.*

Proof. Obvious by II B 2, Theorem 1 and B 4, Theorem 4. If all limbs of X are of the form $B_m I$, then X' is equal to I, otherwise, X' is different from I.

Theorem 2. $\vdash BB \cdot C_1 = C_1 \cdot C_2 \cdot B$.

Proof. First we have by Ax. $(CC)_1$ and II B 2, Theorem 1

$$\begin{aligned}
\vdash C_1 \cdot C_1 &= B_2 I = I \\
\vdash C_2 \cdot C_2 &= B(C_1 \cdot C_1) \quad \text{(II B 4, Thm. 2; II B 3, Def. 1)}. \\
&= B_3 I = I. \quad \text{(Ax. } (CC_1)\text{)}.
\end{aligned}$$

Hence:

$$\begin{aligned}
\vdash BB \cdot C_1 &= C_1 \cdot C_2 \cdot C_2 \cdot C_1 \cdot BB \cdot C_1 \\
&= C_1 \cdot C_2 \cdot B \cdot C_1 \cdot C_1 \quad \text{(Ax. } (BC)\text{)}, \\
&= C_1 \cdot C_2 \cdot B. \quad \text{qed.}
\end{aligned}$$

§2. *The commutative law.*

Convention 1. An equation of the form

$$\vdash C_1 B_{m+1} X = BX \cdot B_n$$

is called a *commutative law* for X, because it allows a certain kind of transposition of X with other entities. Some of the axioms are of this form; I have called these commutative axioms.

Theorem 1. *When X is an entity for which*

$$\vdash CB_{m+1}X = BX \cdot B_n,$$

then for any entity Y

$$\vdash B_m Y \cdot X = X \cdot B_n Y.$$

Proof. By the hypothesis and I D Theorem 6, we have

(1) $\quad\vdash CB_{m+1}XY = (BX \cdot B_n)Y.$
Aber $\quad\vdash CB_{m+1}XY = B_{m+1}YX \quad$ (Reg. C),
$\qquad\qquad\qquad\quad = (B(B_mY))X \quad$ (II B 1, Satz 5),
(2) $\qquad\qquad\quad\; = B_mY \cdot X \quad$ (II B 4, Def. 1),
Auch $\quad\vdash (BX \cdot B_n)Y = BX(B_nY) \quad$ (II B 4, Satz 2),
(3) $\qquad\qquad\qquad = X \cdot B_nY.$

aus (1), (2), (3) wird der Satz bewiesen.

SATZ 2. *Wenn X ein Etwas ist, wofür $\vdash CB_{m+1}X = BX \cdot B_n$; dann*

$$\vdash CB_{m+k+1}X = BX \cdot B_{n+k} \qquad (k = 1, 2, 3, \cdots).$$

Beweis: Wir haben zunächst mit Anwendungen der Eigenschaften der Gleichheit und Definitionen,

$\quad\vdash (CB_{m+1}X) \cdot B_k = (BX \cdot B_n) \cdot B_k$
(1) $\qquad\qquad\qquad\quad = BX \cdot B_{n+k} \qquad$ (II B 4, Sätze 3 und 5).
Aber $\quad\vdash CB_{m+1}X \cdot B_k$
$\qquad\qquad\qquad\quad = B(CB_{m+1}X)B_k \qquad$ (II B 4, Def. 1),
$\qquad\qquad\qquad\quad = BB(CB_{m+1})XB_k \qquad$ (Reg. B),
(2) $\qquad\qquad\qquad = (BB \cdot C)B_{m+1}XB_n \qquad$ (II B 4, Satz 1),
$\qquad\qquad\qquad\quad = (C_1 \cdot C_2 \cdot B)B_{m+1}XB_n \qquad$ (§1, Satz 2),
$\qquad\qquad\qquad\quad = (C_1(C_2(BB_{m+1})))XB_n \qquad$ (II B 4, Sätze 1 und 3),
$\qquad\qquad\qquad\quad = C_2(BB_{m+1})B_nX \qquad$ (Reg. C),
$\qquad\qquad\qquad\quad = C_1(BB_{m+1}B_n)X \qquad$ (Def. von C_2; II B 3,
$\qquad\qquad\qquad\qquad\qquad\qquad\qquad\qquad$ Def. 1; Reg. B),
(3) $\qquad\qquad\qquad = CB_{m+k+1}X \qquad$ (II B 4, Def. 1 und Satz 5).

Aus (1) und (3) wird der Satz bewiesen.

SATZ 3. *Wenn X ein Etwas ist, wofür $\vdash CB_{m+1}X = BX \cdot B_n$; dann gilt für ein beliebiges Etwas Y*

(1) $\vdash B_{m+k+h}Y \cdot B_hX = B_hX \cdot B_{n+k+h}Y \quad (k, h = 0, 1, 2, \cdots).$

Beweis: Nach Satz 2 haben wir

$$\vdash CB_{m+k+1}X = BX \cdot B_{n+k}.$$

$\therefore$ nach Satz 1, $\vdash B_{m+k}Y \cdot X = X \cdot B_{n+k}Y.$
Die Behauptung (1) folgt dann aus II B4, Satz 6 und II B1, Satz 5.

SATZ 4. *Wenn X ein beliebiges Etwas ist; dann gilt für $m = 0, 1, 2, \cdots$, $n = 1, 2, \cdots$, und $p \geq m + 1$,*

$$\vdash B_{p+n}Y \cdot B_mB_n = B_mB_n \cdot B_pY.$$

(1) $\vdash CB_{m+1}XY = (BX \cdot B_n)Y.$
But $\vdash CB_{m+1}XY = B_{m+1}YX$ (Rule C),
$= (B(B_mY))X$ (II B 1, Thm. 5),
(2) $= B_mY \cdot X$ (II B 4, Def. 1),
Also $\vdash (BX \cdot B_n)Y = BX(B_nY)$ (II B 4, Thm. 1),
(3) $= X \cdot B_nY.$

The theorem is proved by (1), (2), (3).

Theorem 2. *If X is an entity where $\vdash CB_{m+1}X = BX \cdot B_n$; then*

$$\vdash CB_{m+k+1}X = BX \cdot B_{n+k} \qquad (k = 1, 2, 3, \cdots).$$

Proof. We first have by application of the properties of equality and definitions,

(1) $\vdash (CB_{m+1}X) \cdot B_k = (BX \cdot B_n) \cdot B_k$
$= BX \cdot B_{n+k}$ (II B 4, Thms. 3, 5).
But $\vdash CB_{m+1}X \cdot B_k$
$= B(CB_{m+1}X)B_k$ (II B 4, Def. 1),
$= BB(CB_{m+1})XB_k$ (Rule B),
(2) $= (BB \cdot C)B_{m+1}XB_k$ (II B 4, Thm. 1),
$= (C_1 \cdot C_2 \cdot B)B_{m+1}XB_k$ (§1, Thm. 2),
$= (C_1(C_2(BB_{m+1})))XB_k$ (II B 4, Thm. 1 & 3),
$= C_2(BB_{m+1})B_kX$ (Rule C),
$= C_1(BB_{m+1}B_k)X$ (Def. of C_2; II B 3, Def. 1; rule B),
(3) $= CB_{m+k+1}X$ (II B 4, Def. 1, Th. 5).

The theorem is now proven by (1) and (3).

Theorem 3. *If X is an entity for which $\vdash CB_{m+1}X = BX \cdot B_n$; then for any entity Y we have*
(1) $\vdash B_{m+k+h}Y \cdot B_hX = B_hX \cdot B_{n+k+h}Y$ $(k, h = 0, 1, 2, \cdots)$.

Proof. By Theorem 2 we have

$$\vdash CB_{m+k+1}X = BX \cdot B_{n+k}.$$

∴ by Theorem 1, $\vdash B_{m+k}Y \cdot X = X \cdot B_{n+k}Y.$

Hence assertion (1) follows from II B 4, Theorem 6 and II B 1, Theorem 5.

Theorem 4. *If Y is an arbitrary entity; then for $m = 0, 1, 2, \cdots$, $n = 1, 2, \cdots$, and $p \geqq m + 1$,*

$$\vdash B_{p+n}Y \cdot B_mB_n = B_mB_n \cdot B_pY.$$

Beweis: Für $n = 1$ folgt der Satz aus Ax. B und Satz 3. Ist der Satz für ein gegebenes n angenommen, so wird er folgendermassen für $n+1$ bewiesen:

$$\begin{aligned}
\vdash B_{p+n+1}Y \cdot B_m B_{n+1} &= B_{p+n+1}Y \cdot B_m B_n \cdot B_m B && \text{(II B 4, Satz 7)}, \\
&= B_m B_n \cdot B_{p+1}Y \cdot B_m B && \text{(Voraussetzung)}, \\
&= B_m B_n \cdot B_m B \cdot B_p Y && \text{(dieser Satz für } n=1\text{)}, \\
&= B_m B_{n+1} \cdot B_p Y && \text{(II B 4, Satz 7)}.
\end{aligned}$$

SATZ 5. *Wenn X ein beliebiges Etwas ist; dann sind die folgenden Gteichungen beweisbar,*

a) $\vdash B_m Y \cdot C_p = C_p \cdot B_m Y$, wenn $m \geq p \geq 1$ gilt,
b) $\vdash B_m Y \cdot W_p = W_p \cdot B_{m+1} Y$, wenn $m \geq p \geq 1$ gilt,
c) $\vdash B_m Y \cdot K_p = K_p \cdot B_{m-1} Y$, wenn $m \geq p \geq 1$ gilt.

Beweis: Diese Gleichungen folgen aus Satz 3, den Axiomen C, W und K und den Definitionen von II B 3.

SATZ 6. *Das Axiom I_1 lässt sich aus den übrigen kombinatorischen Axiomen beweisen.*

Beweis:
$$\begin{aligned}
&\vdash CBI \\
&= CB(WK) && \text{(I C, Def. 3)}, \\
&= B(CB)WK && \text{(Reg. } B\text{)}, \\
&= (B \cdot C)BWK && \text{(II B 4, Satz 1)}, \\
&= (C_2 \cdot C_1 \cdot BB)BWK && \text{(Ax. } (BC)\text{)}, \\
&= C_2(C_1(BBB))WK && \text{(II B 4, Satz 1)}, \\
&= C_1(C_1 B_2 W)K && \text{(Def. von } C_2\text{; Reg. } B\text{; II B 1, Def. 1)}, \\
&= C_1(B_2 W B_2)K && \text{(Ax. } W\text{; II B 4, Satz 1; II B 1, Satz 5)}, \\
&= B_2 C_1 B_2 W B_2 K && \text{(II B 1, Satz 3)}, \\
&= C_1 B_3 C_1 W B_2 K && \text{(Ax. } C\text{; II B 4, Satz 1; II B 1, Satz 5)}, \\
&= B_3 W C_1 B_2 K && \text{(Reg. } C\text{)}, \\
&= BW(C_1 B_2 K) && \text{(II B 1, Satz 2)}, \\
&= BW(BK \cdot I) && \text{(Ax. } K\text{)}, \\
&= W \cdot K_2 \cdot I && \text{(II B 4, Def. 1; Def. von } K_2\text{)}, \\
&= W \cdot C_1 \cdot K \cdot I && \text{(Ax. } (CK)\text{)}, \\
&= W \cdot K \cdot I && \text{(Ax. } (WC)\text{)}, \\
&= BI \cdot I, \text{ w.z.b.w.} && \text{(Ax. } (WK)\text{)}.
\end{aligned}$$

§3. *Umformung in die Form $\Omega \cdot \mathfrak{B}$.*

SATZ 1. *Jedes $\mathfrak{B}$ kann entweder in I oder in die Normalform von II C3, Satz 3, nämlich*

Proof. For $n = 1$ this follows by Ax. B and Theorem 3. If we assume the theorem for a given n, then it will be proven for $n + 1$ as follows:

$$\begin{aligned}
\vdash B_{p+n+1}Y \cdot B_m B_{n+1} &= B_{p+n+1}Y \cdot B_m B_n \cdot B_m && \text{(II B 4, Thm. 7)}, \\
&= B_m B_n \cdot B_{p+1}Y \cdot B_m B && \text{(hypothesis)}, \\
&= B_m B_n \cdot B_m B \cdot B_p Y && \text{(this thm for } n = 1\text{)}, \\
&= B_m B_{n+1} \cdot B_p Y && \text{(II B 4, Thm. 7)}.
\end{aligned}$$

Theorem 5. *If Y is an arbitrary entity; then the following equations are provable,*

a) $\vdash B_m Y \cdot C_p = C_p \cdot B_m Y,$ if $m > p \geq 1$,
b) $\vdash B_m Y \cdot W_p = W_p \cdot B_{m+1} Y,$ if $m \geq p \geq 1$,
c) $\vdash B_m Y \cdot K_p = K_p \cdot B_{m-1} Y,$ if $m \geq p \geq 1$.

Proof. These equations follow from Theorem 3, axioms C, W and K and the definitions from II B 3.

Theorem 6. *Axiom I_1 can be proven from the other combinatory axioms.*

Proof.
$$\begin{aligned}
\vdash\; & CBI \\
=\; & CB(WK) && \text{(I C, Def. 3)}, \\
=\; & B(CB)WK && \text{(rule B)}, \\
=\; & (B \cdot C)BWK && \text{(II B 4, Thm. 1)}, \\
=\; & (C_2 \cdot C_1 \cdot BB)BWK && \text{(Ax. (BC))}, \\
=\; & C_2(C_1(BBB))WK && \text{(II B 4, Thm. 1)}, \\
=\; & C_1(C_1 B_2 W)K && \text{(Def. of } C_2\text{; rule B; II B 1, Def. 1)}, \\
=\; & C_1(B_2 W B_2)K && \text{(Ax. W; II B 4, Thm. 1; II B 1, Thm. 5)}, \\
=\; & B_2 C_1 B_2 W B_2 K && \text{(II B 1, Thm. 3)}, \\
=\; & C_1 B_3 C_1 W B_2 K && \text{(Ax. C; II B 4, Thm. 1; II B 1, Thm. 5)}, \\
=\; & B_3 W C_1 B_2 K && \text{(rule C)}, \\
=\; & BW(C_1 B_2 K) && \text{(II B 1, Thm. 2)}, \\
=\; & BW(BK \cdot I) && \text{(Ax. K)}, \\
=\; & W \cdot K_2 \cdot I && \text{(II B 4, Def. 1; Def. of } K_2\text{)}, \\
=\; & W \cdot C_1 \cdot K \cdot I && \text{(Ax. (CK))}, \\
=\; & W \cdot K \cdot I && \text{(Ax. (WC))}, \\
=\; & BI \cdot I, \text{ qed.} && \text{(Ax. (WK))}.
\end{aligned}$$

§3. *Conversion into the form* $\Omega \cdot \mathfrak{B}$.

Theorem 1. *Every $\mathfrak{B}$ can be converted into either I or the normal form of II C3, Theorem 3, namely*

(1) $$B_{m_q}B_{n_q} \cdot B_{m_{q-1}}B_{n_{q-1}} \cdots B_{m_2}B_{n_2} \cdot B_{m_1}B_{n_1},$$
wo
(2) $$m_1 < m_2 < \cdots < m_q.$$
gilt, umgeformt werden.

Beweis: Nach Festsetzung 4 und §1, Satz 1 kann jedes $\mathfrak{B}$ entweder in I oder in die Form (1) umgeformt werden. Es bleibt nur zu beweisen, dass das betreffende $\mathfrak{B}$ im letzten Fall so umgeformt werden kann, dass auch (2) gilt. Ich beschränke mich auf solche $\mathfrak{B}$s.

Aus §2, Satz 4 und II B4, Satz 7 haben wir

(3) $\vdash B_m B_n \cdot B_p B_q = B_{n+p} B_q \cdot B_m B_n,$ wenn $p > m$ gilt.
(4) $\vdash B_m B_n \cdot B_p B_q = B_m B_{n+q},$ wenn $p = m$ gilt.

Nun sei $\mathfrak{B}$ schon in der Normalform, dann kann $B_r B_s \cdot \mathfrak{B}$ in die Normalform umgeformt werden. In der Tat sei

$$r < m_t, m_{t+1}, \cdots, m_q, \text{ und entwerder } t = 1 \text{ oder } m_{t-1} \leqq r;$$

Dann ist nach (3)
$$B_r B_s \cdot \mathfrak{B} = B_{m_q+s}B_{n_q} \cdot B_{m_{q-1}+s}B_{n_{q-1}} \cdots B_{m_t+s}B_{n_t} \cdot B_r B_s$$
$$\cdots B_{m_{t-1}}B_{n_{t-1}} \cdots B_{m_1}B_{n_1},$$
wo natürlich, wenn $t = 1$ gilt, die Glieder rechts von $B_r B_s$ an der rechten Seite nicht da sind. Wenn $t = 1$ oder $r > m_{t-1}$ gilt, ist die rechte Seite der eben Geschrieben schon in der Normalform. Sonst kann $B_r B_s$ mit sein rechtsstehenden Nachbarn nach (4) verschmolzen werden, und der neue Ausdruck wird in der Normalform sein.

Nun sei $\mathfrak{B}$ ein beliebiger Ausdruck der Form (1). $\mathfrak{B}_r$, sei $(r = 1, 2, \cdots)$ das Produkt der r rechtsstehenden Glieder von $\mathfrak{B}$. $\mathfrak{B}_1$ ist schon in der Nonnalform. Wenn $\mathfrak{B}_r$ in die Normalform umgeformt werden kann, so kann $\mathfrak{B}_{r+1} \equiv B_{m_{r+1}}B_{n_{r+1}} \cdot \mathfrak{B}_r$ nach dem vorigen Absatz in die Normalform umgeformt werden. Also kann $\mathfrak{B}_q \equiv \mathfrak{B}$ in die Normalform umgeformt werden.

SATZ 2. *Zu jedem $\mathfrak{B}$ und C_p gibt es ein $\mathfrak{B}'$ und ein $\mathfrak{C}'$ derart, dass*
$$\vdash \mathfrak{B} \cdot C_p = \mathfrak{C}' \cdot \mathfrak{B}'.$$

Beweis: Für $\mathfrak{B} \equiv I$, klar.
Zunächst sei $\mathfrak{B} \equiv B_m B$. Dann unterscheiden wir vier Fälle:
Fall 1: $p > m + 1$. Dann

$$\vdash B_m B \cdot C_p = C_{p+1} \cdot B_m B \qquad \text{(§2, Satz 4).}$$

Fall 2: $p = m + 1$. Dann

(1) $$B_{m_q}B_{n_q} \cdot B_{m_{q-1}}B_{n_{q-1}} \cdot \cdots \cdot B_{m_2}B_{n_2} \cdot B_{m_1}B_{n_1},$$
where
(2) $$m_1 < m_2 < \cdots < m_q.$$

Proof. By §1 Convention 4 and Theorem 1, every $\mathfrak{B}$ can be converted into either I or the form (1). It remains to show that in the latter case the $\mathfrak{B}$ in question can be converted so that (2) also holds. I restrict myself to such $\mathfrak{B}$s.

By §2, Theorem 4 and II B 4, Theorem 7, we have

(3) $\vdash B_mB_n \cdot B_pB_q = B_{n+p}B_q \cdot B_mB_n,$ if $p > m$.
(4) $\vdash B_mB_n \cdot B_pB_q = B_mB_{n+q},$ if $p = m$.

Now let $\mathfrak{B}$ be already in normal form; then $B_rB_s \cdot \mathfrak{B}$ can be converted into a normal form. In fact, let

$$r < m_t, m_{t+1}, \cdots, m_q, \text{ and either } t = 1 \text{ or } m_{t-1} \leqq r;$$

then by (3)

$$B_rB_s \cdot \mathfrak{B} = B_{m_q+s}B_{n_q} \cdot B_{m_{q-1}+s}B_{n_{q-1}} \cdot \cdots \cdot B_{m_t+s}B_{n_t} \cdot B_rB_s \\ \cdot B_{m_t-1}B_{n_t-1} \cdot \cdots \cdot B_{m_1}B_{n_1},$$

where naturally, if $t = 1$, there are no limbs to the right of B_rB_s. If $t = 1$, or $r > m_{t-1}$, then the right side written above is already in normal form. Otherwise, by (4), B_rR_s can be fused with its neighbours on the right, and the new expression will be in normal form.

Now let $\mathfrak{B}$ be an arbitrary expression in the form (1). Let $\mathfrak{B}_r$ (for $r = 1, 2, \cdots$) be the product of r limbs on the right in $\mathfrak{B}$. $\mathfrak{B}_1$ is already in normal form. If $\mathfrak{B}_r$ is already converted into normal form, then by the previous paragraph, $\mathfrak{B}_{r+1} \equiv B_{m_{r+1}}B_{n_{r+1}} \cdot \mathfrak{B}_r$ can be converted into normal form. Hence $\mathfrak{B}_q \equiv \mathfrak{B}$ can be converted into normal form.

Theorem 2. *For every $\mathfrak{B}$ and C_p there are $\mathfrak{B}'$ and $\mathfrak{C}'$ such that*

$$\vdash \mathfrak{B} \cdot C_p = \mathfrak{C}' \cdot \mathfrak{B}'.$$

Proof. For $\mathfrak{B} \equiv I$, obvious.

Now assume $\mathfrak{B} \equiv B_mB$. Then we differentiate four cases:

Case 1: $p > m + 1$. Then

$$\vdash B_mB \cdot C_p = C_{p+1} \cdot B_mB \quad (\S2, \text{Thm. } 4).$$

Case 2: $p = m + 1$. Then

$$\vdash B_m B \cdot C_{m+1}$$
$$= B_m(B \cdot C_1) \qquad \text{(II B 3, Satz 1; II B 4, Satz 6)},$$
$$= B_m(C_2 \cdot C_1 \cdot BB) \qquad \text{(Ax. }(BC)),$$
$$= C_{m+2} \cdot C_{m+1} \cdot B_{m+1} B \qquad \text{(II B 3, Satz 1; II B 1, Satz 5;}$$
$$\text{II B 4, Satz 6)}.$$

Fall 3: $p = m > 0$. Dann gilt

$$\vdash B_m B \cdot C_m$$
$$= B_{m-1}(BB \cdot C)$$
$$= B_{m-1}(C_1 \cdot C_2 \cdot B) \qquad (\S 1, \text{Satz 2}),$$
$$= C_m \cdot C_{m+1} \cdot B_{m+1} B \qquad \text{(II B 3, Satz 1; II B 4, Satz 6)}.$$

Fall 4: $p < m$. Dann gilt

$$\vdash B_m B \cdot C_p = C_p \cdot B_m B \qquad (\S 2, \text{Satz 5a}).$$

Also ist der Satz für $\mathfrak{B} \equiv B_m B$ bewiesen. Es folgt durch Induktion, dass es zu einem beliebigen $\mathfrak{C}$ ein $\mathfrak{C}'$ und ein $\mathfrak{B}'$ derart gibt, dass

$$\vdash B_m B \cdot \mathfrak{C} = \mathfrak{C}' \cdot \mathfrak{B}'.$$

Das allgemeinste $\mathfrak{B}$ kann nun entweder in I oder in ein Produkt von N Faktoren der Form $B_m B$ (Satz 1, II B4, Satz 7)* umgeformt werden. Wir können nun ohne Beschränkung der Allgemeinheit $\mathfrak{B}$ als in dieser letzten Form gegeben betrachten. $\mathfrak{B}_M$ sei das Produkt der M rechtsstehenden Faktoren von $\mathfrak{B}$. Wenn der Satz für jedes $\mathfrak{B}_M$ bewiesen ist, so ist er für $\mathfrak{B}$ bewiesen. Aber für $M = 1$ ist er schon im letzten Absatz bewiesen. Für bestimmtes M sei angenommen, dass $\vdash \mathfrak{B}_M \cdot C_p = \mathfrak{C} \cdot \mathfrak{B}'_M$, dann gilt

$$\vdash \mathfrak{B}_{M+1} \cdot C_p = B_m B \cdot \mathfrak{B}_M \cdot C_p$$
$$= B_m B \cdot \mathfrak{C} \cdot \mathfrak{B}'_M$$
$$= \mathfrak{C}' \cdot \mathfrak{B}'' \cdot \mathfrak{B}'_M$$
$$= \mathfrak{C}' \cdot \mathfrak{B}'.$$

Daher folg der Satz durch Induktion für alle $\mathfrak{B}_M$, also auch für $\mathfrak{B}$.

SATZ 3. *Wenn X ein regulärer Kombinator ist, dessen sämtliche Glieder der form $B_m B_n$ oder C_p sind, so kann X in die Form $(\mathfrak{C} \cdot \mathfrak{B})$ umgeformt werden.*

Beweis: Es sei $X \equiv X_1 \cdot X_2 \cdots X_n$ wo die X_i die Glieder von X sind.

Es sei nun angenommen, dass $(X_1 \cdot X_2 \cdots X_q)$ in die Form $(\mathfrak{C}' \cdot \mathfrak{B}')$ umgeformt werden kann; dann gilt, wenn $X_{q+1} \equiv B_m B_n$,

---

*Für das $\mathfrak{B}$ von II C3 (1) Satz 2 ist $N = n_1 + n_2 + n_3 + \cdots + n_q$.

$\vdash B_m B \cdot C_{m+1}$
$= B_m(B \cdot C_1)$ (II B 3, Thm. 1; II B 4, Thm. 6),
$= B_m(C_2 \cdot C_1 \cdot BB)$ (Ax. (BC)),
$= C_{m+2} \cdot C_{m+1} \cdot B_{m+1} B$ (II B 3, Thm. 1; II B 1, Thm. 5; II B 4, Thm. 6).

Case 3: $p = m > 0$. Then
$\vdash B_m B \cdot C_m$
$= B_{m-1}(BB \cdot C)$
$= B_{m-1}(C_1 \cdot C_2 \cdot B)$ (§1, Thm. 2),
$= C_m \cdot C_{m+1} \cdot B_{m-1} B$ (II B 3, Thm. 1; II B 4, Thm. 6).

Case 4: $p < m$. Then
$$\vdash B_m B \cdot C_p = C_p \cdot B_m B \quad (\S 2, \text{Thm. 5a}).$$

Hence the theorem is proven for $\mathfrak{B} \equiv B_m B$. It follows by induction, that for an arbitrary $\mathfrak{C}$, there are $\mathfrak{C}'$ and $\mathfrak{B}'$, such that
$$\vdash B_m B \cdot \mathfrak{C} = \mathfrak{C}' \cdot \mathfrak{B}'.$$

The most general $\mathfrak{B}$ can now be converted into either I or a product of N factors of the form $B_m B$ (Thm. 1, II B 4, Thm. 7).* Without loss of generality we can now consider $\mathfrak{B}$ of this latter form. Let $\mathfrak{B}_M$ be the product of the M righthand factors of $\mathfrak{B}$. If the theorem is proven for every $\mathfrak{B}_M$, then it is proven for $\mathfrak{B}$. But for $M = 1$, it is already proven in the last paragraph. Assume for some M, that $\vdash \mathfrak{B}_M \cdot C_p = \mathfrak{C} \cdot \mathfrak{B}'_M$, then
$$\begin{aligned}
\vdash \mathfrak{B}_{M+1} \cdot C_p &= B_m B \cdot \mathfrak{B}_M \cdot C_p \\
&= B_m B \cdot \mathfrak{C} \cdot \mathfrak{B}'_M \\
&= \mathfrak{C}' \cdot \mathfrak{B}'' \cdot \mathfrak{B}'_M \\
&= \mathfrak{C}' \cdot \mathfrak{B}'.
\end{aligned}$$

Hence the theorem follows by induction for all $\mathfrak{B}_M$, and hence also for $\mathfrak{B}$.

Theorem 3. *If X is a regular combinator, whose limbs are all of the form $B_m B_n$ or C_p, then X can be converted into the form $(\mathfrak{C} \cdot \mathfrak{B})$.*

Proof. Assume $X \equiv X_1 \cdot X_2 \cdot \ldots \cdot X_n$ where these X_is are the limbs of X. Assume now that $(X_1 \cdot X_2 \cdot \ldots \cdot X_q)$ can be converted into the form $(\mathfrak{C}' \cdot \mathfrak{B}')$; then when $X_{q+1} \equiv B_m B_n$ we have

---

*For the $\mathfrak{B}$ of II C3 (1) Theorem 2, $N = n_1 + n_2 + n_3 + \cdots + n_q$.

$$\vdash X_1 \cdot X_2 \cdots X_q \cdot X_{q+1} = \mathfrak{C}' \cdot \mathfrak{B}' \cdot B_m B_n$$
$$= \mathfrak{C}' \cdot \mathfrak{B}'',$$

während, wenn $X_{q+1} \equiv C_p$ ist, gilt

$$\vdash X_1 \cdot X_2 \cdots X_{q+1} = \mathfrak{C}' \cdot \mathfrak{B}' \cdot C_p$$
$$= \mathfrak{C}' \cdot \mathfrak{C}'' \cdot \mathfrak{B}'' \quad \text{(Satz 2)},$$
$$= \mathfrak{C}''' \cdot \mathfrak{B}''.$$

Also ist der Satz durch Induktion auf q für X bewiesen, weil er für $q = 1$ klar ist.

SATZ 4. *Für jedes $\mathfrak{B}$ und W_p gibt es ein Ω und ein $\mathfrak{B}'$ derart, dass*

$$\vdash \mathfrak{B} \cdot W_p = \Omega \cdot \mathfrak{B}'.$$

Beweis: Für $\mathfrak{B} \equiv I$ klar. Ich beschränke mich also auf den Fall $\mathfrak{B} \not\equiv I$.
Zunächst zeige ich, dass es für jedes $m = 0, 1, 2, \cdots, p = 0, 1, 2, \cdots, k = 0, 1, 2, \cdots, p-1$, ein $q > 0$, ein $h < q$ und ein X, deren sämtliche Glieder der Form $B_m B_n$ oder C_p sind, derart gibt, dass

(7) $\quad \vdash B_m B \cdot W_p \cdot W_{p-1} \cdots W_{p-k} = W_q \cdot W_{q-1} \cdots W_{q-h} \cdot X.$

Es sind drei Fälle zu unterscheiden:
Fall 1. $p \leqq m$. Weil nach §2, Satz 5 für alle $r \leqq m$

$$\vdash B_m B \cdot W_r = W_r \cdot B_{m+1} B,$$

so haben wir hier,

$$\vdash B_m B \cdot W_p \cdot W_{p-1} \cdots W_{p-k} = W_p \cdot W_{p-1} \cdots W_{p-k} \cdot B_{m+k+1} B.$$

Fall 2. $p = m+1$. Hier gilt für $k = 0$,

$\vdash B_m B \cdot W_{m+1}$	$= B_m(B \cdot W)$	(II B 4, Satz 6),
	$= B_m(W_2 \cdot W_1 \cdot C_2 \cdot B_2 B \cdot B)$	(Ax. (BW)),
	$= W_{m+2} \cdot W_{m+1} \cdot C_{m+2} \cdot B_{m+2} B \cdot B_m B$	(II B 4, Satz 6)

Für $k \geqq 1$,
$\vdash B_m B \cdot W_{m+1} \cdot W_m \cdots W_{m-k+1}$
$= W_{m+2} \cdot W_{m+1} \cdot C_{m+2} \cdot B_{m+2} B \cdot B_m B \cdot W_m \cdots W_{m-k+1}$
$\hspace{10em}$ (nach dem Falle $k = 0$),
$= W_{m+2} \cdot W_{m+1} \cdot C_{m+2} \cdot W_m \cdot W_{m-1} \cdots W_{m-k+1} \cdot B_{m+k+2} B \cdot B_{m+k} B$
$\hspace{10em}$ (nach Falle 1),
$= W_{m+2} \cdot W_{m+1} \cdot W_m \cdots W_{m-k+1} \cdot C_{m+k+2} \cdot B_{m+k+2} B \cdot B_{m+k} B,$

wobei der letzte Schritt aus §2, Satz 5 und Definition von C_{m+k+2} folgt.

$$\vdash X_1 \cdot X_2 \cdot \cdots \cdot X_q \cdot X_{q+1} = \mathfrak{C}' \cdot \mathfrak{B}' \cdot B_m B_n$$
$$= \mathfrak{C}' \cdot \mathfrak{B}'',$$

whereas, when $X_{q+1} \equiv C_p$, we have

$$\vdash X_1 \cdot X_2 \cdot \cdots \cdot X_q \cdot X_{q+1} = \mathfrak{C}' \cdot \mathfrak{B}' \cdot C_p$$
$$= \mathfrak{C}' \cdot \mathfrak{C}'' \cdot \mathfrak{B}'' \quad \text{(Thm. 2)},$$
$$= \mathfrak{C}''' \cdot \mathfrak{B}''.$$

Hence the theorem is proven for X by induction on q, because it is obvious when $q = 1$.

Theorem 4. *For every $\mathfrak{B}$ and W_p there exist an Ω and a $\mathfrak{B}'$ such that*

$$\vdash \mathfrak{B} \cdot W_p = \Omega \cdot \mathfrak{B}'.$$

Proof. For $\mathfrak{B} \equiv I$, obvious. I limit myself to the case $\mathfrak{B} \not\equiv I$. First I show that for every $m = 0, 1, 2, \cdots$, $p = 0, 1, 2, \cdots$, $k = 0, 1, 2, \cdots, p-1$, there is a $q > 0$, an $h < q$ and an X all of whose limbs are of the form $B_m B_n$ or C_p, such that

(7) $\quad \vdash B_m B \cdot W_p \cdot W_{p-1} \cdot \cdots \cdot W_{p-k} = W_q \cdot W_{q-1} \cdot \cdots \cdot W_{q-h} \cdot X.$

There are three cases to distinguish:

Case 1. $p \leqq m$. Because by §2, Theorem 5 for all $r \leqq m$

$$\vdash B_m B \cdot W_r = W_r \cdot B_{m+1} B,$$

then we have here,

$$\vdash B_m B \cdot W_p \cdot W_{p-1} \cdot \cdots \cdot W_{p-k} = W_p \cdot W_{p-1} \cdot \cdots \cdot W_{p-k} \cdot B_{m+k+1} B.$$

Case 2. $p = m + 1$. Then for $k = 0$,

$$\vdash B_m B \cdot W_{m+1} = B_m(B \cdot W) \qquad \text{(II B 4, Thm. 6)},$$
$$= B_m(W_2 \cdot W_1 \cdot C_2 \cdot B_2 B \cdot B) \qquad \text{(Ax. (BW))},$$
$$= W_{m+2} \cdot W_{m+1} \cdot C_{m+2} \cdot B_{m+2} B \cdot B_m B \qquad \text{(II B 4, Thm. 6)}$$

For $k \geqq 1$,

$$\vdash B_m B \cdot W_{m+1} \cdot W_m \cdot \cdots \cdot W_{m-k+1}$$
$$= W_{m+2} \cdot W_{m+1} \cdot C_{m+2} \cdot B_{m+2} B \cdot B_m B \cdot W_m \cdot \cdots \cdot W_{m-k+1}$$
$$\text{(by the case } k = 0\text{),}$$
$$= W_{m+2} \cdot W_{m+1} \cdot C_{m+2} \cdot W_m \cdot W_{m-1} \cdot \cdots \cdot W_{m-k+1} \cdot B_{m+k+2} B \cdot B_{m+k} B$$
$$\text{(by Case 1),}$$
$$= W_{m+2} \cdot W_{m+1} \cdot W_m \cdot \cdots \cdot W_{m-k+1} \cdot C_{m+k+2} \cdot B_{m+k+2} B \cdot B_{m+k} B,$$

whereby the last step follows by §2, Theorem 5 and the definition of C_{m+k+2}.

Fall 3. $p > m + 1$. Aus §2, Satz 4 folgt, für $k = 0$,
$$\vdash B_m B \cdot W_p = W_{r+1} \cdot B_m B.$$

Also:

$$\vdash B_m B \cdot W_p \cdot W_{p+1} \cdots W_{m+3} \cdot W_{m+2} \cdot W_{m+1} \cdots W_{p-k},$$
$$= W_{p+1} \cdot W_p \cdots W_{m+3} W_{m+2} \cdot B_m B \cdot W_{m+1} \cdots W_{p-k},$$
$$= W_{p+1} \cdot W_p \cdots W_{m+3} \cdot W_{m+2} \cdots W_{p-k} \cdot C_{m+k+2} \cdot B_{m+k+2} \cdot B_{m+k} B$$
(Fall 2).

Also ist (7) bewiesen. Durch den Induktionsprozess, den ich im letzten Absatz des Beweises von Satz 2 benutzt habe, wird die Gleichung bewiesen, die entsteht wenn man in (7) $B_m B$ durch ein beliebiges $\mathfrak{B}$ ersetzt. Wenn man in dieser Gleichung $k = 0$ setzt, so hat man

$$\vdash \mathfrak{B} \cdot W_p = W_q \cdot W_{q-1} \cdots W_{q-h} \cdot X,$$

wo, nach Satz 3
$$\vdash X = \mathfrak{C} \cdot \mathfrak{B}'.$$
Also
$$\vdash \mathfrak{B} \cdot W_p = W_q \cdot W_{q-1} \cdots W_{q-h} \cdot \mathfrak{C} \cdot \mathfrak{B}'$$

$$= \Omega \cdot \mathfrak{B}' \qquad \text{w.z.b.w.,}$$

weil $(W_q \cdot W_{q-1} \cdots W_{q-h} \cdot \mathfrak{C})$ die Definition eines Ω erfüllt.

SATZ 5. *Für jedes $\mathfrak{B}$ und K_p gibt es ein $\mathfrak{K}$ und ein $\mathfrak{B}'$ derart, dass $\vdash \mathfrak{B} \cdot K_p = \mathfrak{K} \cdot \mathfrak{B}'$ oder auch, gibt es ein $\mathfrak{K}$, wofür $\vdash \mathfrak{B} \cdot K_p = \mathfrak{K}$.*

Beweis: Ich beweise den Satz zunächst für den Fall $\mathfrak{B} \equiv B_m B$. Dann gibt es drei Fälle:

Fall 1: $p \leqq m$. Dann gilt nach §2, Satz 5
$$\vdash B_m B \cdot K_p = K_p \cdot B_{m-1} B.$$

Fall 2: $p = m + 1$. Dann folgt aus Ax. (BK) und II B4, Satz 6
$$\vdash B_m B \cdot K_{m+1} = B_m(B \cdot K) = B_m(K_1 \cdot K_1) = K_{m+1} \cdot K_{m+1}.$$

Fall 3: $p > m + 1$. Dann folg aus §2, Satz 4,
$$\vdash B_m B \cdot K_p = K_{p+1} \cdot B_m B.$$

Der Rest des Beweises läuft genau wie in Satz 2.

SATZ 6. *Jeder reguläre Kombinator lässt sich in die Form $(\Omega \cdot \mathfrak{B})$ umformen.*

Beweis: X sei ein regulärer Kombinator und $X_1, X_2, \cdots X_q$ seien seine Glieder, so dass
$$X \equiv X_1 \cdot X_2 \cdots X_n.$$

Case 3. $p > m+1$. By §2, Theorem 4, we have for $k = 0$,
$$\vdash B_m B \cdot W_p = W_{p+1} \cdot B_m B.$$
Hence:
$$\vdash B_m B \cdot W_p \cdot W_{p+1} \cdots W_{m+3} \cdot W_{m+2} \cdot W_{m+1} \cdots W_{p-k},$$
$$= W_{p+1} \cdot W_p \cdots W_{m+3} \cdot B_m B \cdot W_{m+1} \cdots W_{p-k},$$
$$= W_{p+1} \cdot W_p \cdots W_{m+3} \cdot W_{m+2} \cdots W_{p-k} \cdot C_{m+k+2} \cdot B_{m+k+2} \cdot B_{m+k} B$$
(Case 2).

Hence (7) is proven. By the induction process, that I have used in the last paragraph of the proof of Theorem 2, the equation is now proven, which results when one replaces $B_m B$ in (7) by an arbitrary $\mathfrak{B}$. If one takes $k = 0$ in this equation, then one has
$$\mathfrak{B} \cdot W_p = W_q \cdot W_{q-1} \cdots W_{q-h} \cdot X,$$
where by Theorem 3, $\qquad \vdash X = \mathfrak{C} \cdot \mathfrak{B}'.$

Hence $\qquad \vdash \mathfrak{B} \cdot W_p = W_q \cdot W_{q-1} \cdots W_{q-h} \cdot \mathfrak{C} \cdot \mathfrak{B}'$
$$= \Omega \cdot \mathfrak{B}' \text{ qed.},$$
because $(W_q \cdot W_{q-1} \cdots W_{q-h} \cdot \mathfrak{C})$ satisfies the definition of Ω.

Theorem 5. *For every $\mathfrak{B}$ and K_p there exist a $\mathfrak{K}$ and a $\mathfrak{B}'$ such that $\vdash \mathfrak{B} \cdot K_p = \mathfrak{K} \cdot \mathfrak{B}'$ or there is a $\mathfrak{K}$ for which $\vdash \mathfrak{B} \cdot K_p = \mathfrak{K}$.*

Proof. I prove the theorem first for the case $\mathfrak{B} \equiv B_m B$. Then there are three cases:

Case 1: $p \leqq m$. Then by §2, Theorem 5
$$\vdash B_m B \cdot K_p = K_p \cdot B_{m-1} B.$$

Case 2: $p = m + 1$. Then by Ax. (BK) and II B 4, Theorem 6
$$\vdash B_m B \cdot K_{m+1} = B_m(B \cdot K) = B_m(K_1 \cdot K_1) = K_{m+1} \cdot K_{m+1}.$$

Case 3: $p > m + 1$. Then by §2, Theorem 4,
$$\vdash B_m B \cdot K_p = K_{p+1} \cdot B_m B.$$

The rest of the proof goes exactly as in Theorem 2.

Theorem 6. *Every regular combinator can be converted into the form $(\Omega \cdot \mathfrak{B})$.*

Proof. Let X be a regular combinator and $X_1, X_2, \cdots, X_q$ be its limbs, so that
$$X \equiv X_1 \cdot X_2 \cdots X_q.$$

Der Satz ist sicher wahr für X_1. Nehmen wir an, er ist für den Kombinator $(X_1 \cdot X_2 \cdots X_q)$ wahr, dann werde ich ihn für $(X_1 \cdot X_2 \cdots X_{q+1})$. beweisen. In der Tat sei

$$\vdash X_1 \cdot X_2 \cdots X_q = \Omega' \cdot \mathfrak{B}'.$$

Dann ist X_{q+1} entweder $B_m B_n$, C_p, W_p oder K_p.* Im ersten Fall ist das zu Beweisende klar, wenn wir $\Omega \equiv \Omega'$, $\mathfrak{B} \equiv \mathfrak{B}' \cdot B_m B_n$ setzen. In anderen Fällen wissen wir aus den Sätzen 2, 4 und 5, dass es ein Ω'' und ein $\mathfrak{B}''$ gibt, wofür $\vdash \mathfrak{B}' \cdot X_{q+1} = \Omega'' \cdot \mathfrak{B}''$ gilt, also

$$\begin{aligned}\vdash X_1 \cdot X_2 \cdots X_{q+1} &= \Omega' \cdot \mathfrak{B}' \cdot X_{q+1} \\ &= \Omega' \cdot \Omega'' \cdot \mathfrak{B}'' \\ &= \Omega \cdot \mathfrak{B},\end{aligned}$$

wenn wir $\Omega \equiv \Omega' \cdot \Omega''$, $\mathfrak{B} \equiv \mathfrak{B}''$ definieren.

§4. Die Umformung $\Omega = \mathfrak{K} \cdot \mathfrak{M}$.

SATZ 1. *Jedes Ω kann in die Form $(\mathfrak{K} \cdot \mathfrak{M})$ umgeformt werden.*

Beweis: Es genügt zu zeigen, dass jeder Kombinator der Form $(\mathfrak{M} \cdot K_p)$ in die betreffende Form übergeführt werden kann, denn das allgemeinste Ω enthält entweder kein K–und dann ist der Satz klar ($\mathfrak{K} \equiv I$)–, oder es kann in die Form

$$(\mathfrak{M}_1 \cdot K_{p_1} \cdot \mathfrak{M}_2 \cdot K_{p_2} \cdots \mathfrak{M}_k \cdot K_{p_k} \cdot \mathfrak{M}_{k+1})$$

umgeformt werden, wo einzelne $\mathfrak{M}_i \equiv I$ können. (In der Tat folgt dies durch Einschaltungen von gewissen I's welche durch II B 4, Satz 4 erlaubt sind). Dann wird durch Wiederholung des Prozesses wodurch $(\mathfrak{M} \cdot K_p)$ in die Form des Satzes umgeformt wird, der ganze Ausdruck in diese Form gebracht.

Weiter genügt es zu beweisen, dass $W_m \cdot K_p$ in die Form $K_r \cdot W_s$ bzw. I und $C_m \cdot K_p$ in die Form $K_r \cdot C_s$ bzw. K_r umgeformt werden können. Denn wenn diese Behauptungen bewiesen sind, so folgt daraus, dass die einzelnen Glieder eines $\mathfrak{M}$ eins nach dem andern über die K's übertragen oder mit ihnen verschmolzen werden können.

Die Behandlung von $(C_m \cdot K_p)$ gibt vier Fälle:
Fall 1: $p \leq m - 1$. Dann

$$\vdash C_m \cdot K_p = K_p \cdot C_{m-1} \qquad (\S 2, \text{Satz 5c, II B 3}).$$

---

*Wir können natürlich annehmen, dass keine Glieder der Form $B_m I$ vorkommen, weil der Satz für $X \equiv I$ klar ist (s. §1, Satz 1).

The statement is certainly true for X_1. Assume it is true for the combinator $(X_1 \cdot X_2 \cdots\cdots X_q)$, then I will show it for $(X_1 \cdot X_2 \cdots\cdots X_{q+1})$. In fact, let

$$\vdash X_1 \cdot X_2 \cdots\cdots X_q = \Omega' \cdot \mathfrak{B}'.$$

Then X_{q+1} is either $B_m B_n$, C_p, W_p or K_p.* In the first case the proof is obvious if we set $\Omega \equiv \Omega'$, $\mathfrak{B} \equiv \mathfrak{B}' \cdot B_m B_n$. In the other cases we know by Theorems 2, 4 and 5 that there exist an Ω'' and a $\mathfrak{B}''$ for which $\vdash \mathfrak{B}' \cdot X_{q+1} = \Omega'' \cdot \mathfrak{B}''$ holds; hence

$$\begin{aligned}\vdash X_1 \cdot X_2 \cdots\cdots X_{q+1} &= \Omega' \cdot \mathfrak{B}' \cdot X_{q+1} \\ &= \Omega' \cdot \Omega'' \cdot \mathfrak{B}'' \\ &= \Omega \cdot \mathfrak{B},\end{aligned}$$

where we define $\Omega \equiv \Omega' \cdot \Omega''$ and $\mathfrak{B} \equiv \mathfrak{B}''$.

§4. *The Conversion* $\Omega = \mathfrak{K} \cdot \mathfrak{M}$.

Theorem 1. *Every Ω can be converted into the form $(\mathfrak{K} \cdot \mathfrak{M})$.*

Proof. It suffices to show that every combinator in the form $(\mathfrak{M} \cdot K_p)$ can be transformed into the form in question, because either the most general Ω does not contain any K –and then the theorem is obvious $(\mathfrak{K} \equiv I)$–, or it can be converted into the form

$$(\mathfrak{M}_1 \cdot K_{p_1} \cdot \mathfrak{M}_2 \cdot K_{p_2} \cdots\cdots \mathfrak{M}_k \cdot K_{p_k} \cdot \mathfrak{M}_{k+1})$$

where it is possible to have $\mathfrak{M}_i \equiv I$. (In fact this follows from the insertions of some I's, which are possible by II B 4, Theorem 4). Then by repetition of the process by which $(\mathfrak{M} \cdot K_p)$ is converted into the form of the theorem, the entire expression can be brought into this form.

Further it suffices to prove that $W_m \cdot K_p$ can be converted to the form $K_r \cdot W_s$ or I and $C_m \cdot K_p$ can be converted to the form $K_r \cdot C_s$ or K_r. Because when these assertions are proven, it will follow that the individual limbs of an $\mathfrak{M}$ can either be transmitted one after the other over the K's or can be fused with them.

The treatment of $(C_m \cdot K_p)$ has four cases:

Case 1: $p \leqq m - 1$. Then

$$\vdash C_m \cdot K_p = K_p \cdot C_{m-1} \qquad \text{§2, Thm. 5c, II B 3).}$$

---

*We can naturally assume that no limbs of the form $B_m I$ appear, because the theorem for $X \equiv I$ is obvious (see §1, Theorem 1).

Fall 2. $p = m$. Dann

$$\begin{aligned}
\vdash C_m \cdot K_m &= B_{m-1}(C_1 \cdot K_1) && \text{(II B 4, Satz 6; II B 3)},\\
&= B_{m-1} K_2 && \text{(Ax. }(CK)),\\
&= K_{m+1} && \text{(II B 3, Satz 5).}
\end{aligned}$$

Fall 3. $p = m + 1$. Nach Fall 2 folgt

$$\begin{aligned}
\vdash C_m &\cdot K_{m+1}\\
&= C_m \cdot C_m \cdot K_m\\
&= B_m(C_1 \cdot C_1) \cdot K_m && \text{(II B 3, Satz 1)},\\
&= B_{m+2} I \cdot K_m && \text{(Ax. }(CC)_1\text{, II B 1, Satz 5)},\\
&= K_m && \text{(II B 2, Satz 1, und II B 4, Satz 4).}
\end{aligned}$$

Fall 4: $p > m + 1$. Dann nach §2, Satz 5,

$$\vdash C_m \cdot K_p = K_p \cdot C_m.$$

Die Behandlung für $(W_m \cdot K_p)$ gibt drei Fälle:

Fall 1: $p \leqq m - 1$.

$$\vdash W_m \cdot K_p = K_p \cdot W_{m-1} \qquad \text{(§2, Satz 5c; II B 3, Sätze 3 u. 5).}$$

Fall 2: $p = m$.

$$\begin{aligned}
\vdash W_m &\cdot K_m\\
&= B_{m-1}(W_1 \cdot K_1) && \text{(II B 4, Satz 6; II B 3, Sätze 3 u. 5))},\\
&= B_m I && \text{(Ax. }(WK)\text{; II B 1, Satz 5)},\\
&= I && \text{(II B 2, Satz 1).}
\end{aligned}$$

Fall 3: $p > m$. Dann

$$\vdash W_m \cdot K_p = K_{p-1} \cdot W_m \quad \text{(§2, Satz 5b; II B 3, Satz 3 u. 5).}$$

Damit ist der Satz vollständig bewiesen.

SATZ 2. *Jedes $\mathfrak{K}$ kann entwerder in I oder in die Normalform von II C 4, Satz 3, nämlich*

(1) $$(K_{h_p} \cdot K_{h_{p-1}} \cdots K_{h_2} \cdot K_{h_1})$$

wo

(2) $$h_1 < h_2 < \cdots < h_p.$$

sind, umgeformt werden.

Beweis: Nach §1, Festsetzung 4 und §1, Satz 1 kann $\mathfrak{K}$ entweder auf I oder auf die Form (1) gebracht werden. Aus §2, Satz 5 folgt

(3) $$\vdash K_m \cdot K_p = K_{p+1} \cdot K_m, \quad \text{wenn } p \geqq m.$$

Wenn es in dem betreffenden Ausdruck zwei benachbarte K's etwa K_{h_s} und $K_{h_{s-1}}$ gibt, wofür $h_{s-1} \geqq h_s$ ist, so kann eine gewisse Vertauschung stattfinden.

Case 2: $p = m$. Then
$$\begin{aligned} \vdash C_m \cdot K_m &= B_{m-1}(C_1 \cdot K_1) &&\text{(II B 4, Thm. 6; II B 3)}, \\ &= B_{m-1}K_2 &&\text{(Ax. } (CK)), \\ &= K_{m+1} &&\text{(II B 3, Thm. 5)}. \end{aligned}$$

Case 3: $p = m + 1$. By Case 2 we have
$$\begin{aligned} \vdash C_m &\cdot K_{m+1} \\ &= C_m \cdot C_m \cdot K_m \\ &= B_{m-1}(C_1 \cdot C_1) \cdot K_m &&\text{(II B 3, Thm. 1)}, \\ &= B_{m+1}I \cdot K_m &&\text{(Ax. } (CC)_1, \text{II B 1, Thm. 5)}, \\ &= K_m &&\text{(II B 2, Thm. 1, and II B 4, Thm. 4)}. \end{aligned}$$

Case 4: $p > m + 1$. Then by §2, Theorem 5,
$$\vdash C_m \cdot K_p = K_p \cdot C_m.$$

The treatment of $(W_m \cdot K_p)$ has three cases:
Case 1: $p \leq m - 1$. Then
$$\vdash W_m \cdot K_p = K_p \cdot W_{m-1} \qquad (\text{§2, Thm. 5c; II B 3, Thm. 3 \& 5}).$$

Case 2: $p = m$. Then
$$\begin{aligned} \vdash W_m &\cdot K_m \\ &= B_{m-1}(W_1 \cdot K_1) &&\text{(II B 4, Thm. 6; II B 3, Thm. 3 \& 5))}, \\ &= B_m I &&\text{(Ax. } (WK); \text{II B 1, Thm. 5)}, \\ &= I &&\text{(II B 2, Thm. 1)}. \end{aligned}$$

Case 3: $p > m$. Then
$$\vdash W_m \cdot K_p = K_{p-1} \cdot W_m \quad (\text{§2, Thm. 5b; II B 3, Thm. 3 \& 5}).$$

So the theorem is completely proven.

Theorem 2. *Every $\mathfrak{K}$ can be either be converted into I or into the normal form of II C4, Theorem 3, namely*

(1) $\qquad\qquad (K_{h_p} \cdot K_{h_{p-1}} \cdots\cdots K_{h_2} \cdot K_{h_1})$
where
(2) $\qquad\qquad h_1 < h_2 < \cdots < h_p.$

Proof. By §1, Convention 4 and §1, Theorem 1, $\mathfrak{K}$ can be transformed into either I or into the form (1). By §2, Theorem 5, it follows that

(3) $\qquad\qquad \vdash K_m \cdot K_p = K_{p+1} \cdot K_m, \quad \text{if } p \geq m.$

When in the expression in question there are two adjacent K's like K_{h_s} and $K_{h_{s-1}}$ for $h_{s-1} \geq h_s$, a certain exchange can take place.

Nach einer gewissen Anzahl von Vertauschungen nach (3) wird der Ausdruck auf eine Form, wo (2) zutrifft, gebracht. Der genaue Beweis verläuft hier wie im §3, Satz 1.

§5. *Die Normalform für* $\mathfrak{M}$.

In meiner oben erwähnten Abhandlung habe ich schon bewiesen, dass aus gewissen Axiomen (besser Axiomenschemen, wovon einige unendlich viele Axiome enthalten) die folgenden sich schliessen lassen: 1) jedes $\mathfrak{M}$ kann in die Normalform umgeformt werden, 2) wenn $\mathfrak{M}_1$ und $\mathfrak{M}_2$ derselben Folge entsprechen, so folgt $\vdash \mathfrak{M}_1 = \mathfrak{M}_2$. Um diese Ergebnis unserer Theorie zu sichern, genügt es zu beweisen, dass die dort gegebenen Axiomen, und auch die Definitionen von $W_2, W_3 \cdots$ aus unserem Grundgerüst ableitbar sind.

SATZ 1. $\vdash C_m \cdot C_m = I$ $(m = 1, 2, 3, \cdots)$.

Beweis: Nach Definition von C_m und II B 4, Satz 6 gilt

$$
\begin{aligned}
\vdash C_m \cdot C_m &= B_{m-1}(C_1 \cdot C_1) \\
&= B_{m-1}(B_2 I) \quad \text{(Ax. } (CC)_1\text{)}, \\
&= I \quad \text{(II B 1, Satz 5; II B 2, Satz 1)}.
\end{aligned}
$$

SATZ 2. $\vdash C_m \cdot C_{m+1} \cdot C_m = C_{m+1} \cdot C_m \cdot C_{m+1}$ $(m = 1, 2, \cdots)$.

Beweis:

$$
\begin{aligned}
&\vdash C_m \cdot C_{m+1} \cdot C_m \\
&= B_{m-1}(C_1 \cdot C_2 \cdot C_1) \quad \text{(II B 4, Sätze 3 und 6)}, \\
&= B_{m-1}(C_2 \cdot C_1 \cdot C_2) \quad \text{(Ax. } (CC)_2\text{)}, \\
&= C_{m+1} \cdot C_m \cdot C_{m+1}.
\end{aligned}
$$

SATZ 3. $\vdash C_m \cdot C_{m+j} = C_{m+j} \cdot C_m$, wenn $j > 1$, $(m = 1, 2, \cdots)$.

Beweis: Folgt aus §2, Satz 5, wenn wir C_{j-1} für Y in die Gleichung a) setzen.

SATZ 4. $\vdash C_m \cdot W = W \cdot C_{m+1}$ $(m = 2, 3, 4, \cdots)$.

Beweis: gleich aus §2, Satz 5b.

SATZ 5. $\vdash W_m \cdot W_n = W_n \cdot W_{m+1}$ $(m \geqq n = 1, 2, 3, \cdots)$.

Beweis: Für $m = n$,

$$
\begin{aligned}
&\vdash W_m \cdot W_m \\
&= B_{m-1}(W_1 \cdot W_1) \quad \text{(II B 3, Def. 2; II B 4, Satz 6)}, \\
&= B_{m-1}(W_1 \cdot W_2) \quad \text{(Ax. } (WW)\text{)}, \\
&= W_m \cdot W_{m+1} \quad \text{(II B 3, Def. 2; II B 4, Satz 6)}.
\end{aligned}
$$

Für $m > n$ folg der Satz aus §2, Satz 5b.

After a certain number of exchanges by (3) the expression can be brought into a form where (2) holds. The precise proof goes here as in §3, Theorem 1.

§5. *The normal form for* $\mathfrak{M}$.

In my above mentioned paper I have already proven that from certain axioms (better axiom schemes, which include infinitely many axioms) the following can be inferred: 1) every $\mathfrak{M}$ can be converted into a normal form, 2) If $\mathfrak{M}_1$ and $\mathfrak{M}_2$ conform to the same sequence then $\mathfrak{M}_1 = \mathfrak{M}_2$. To secure these results for our theory, it is enough to prove that the axioms given there and also the definitions of $W_1, W_2, \cdots$ are derivable from our primitive frame.

Theorem 1. $\vdash C_m \cdot C_m = I \quad (m = 1, 2, 3, \cdots)$.

Proof. By the definition of C_m and II B 4, Theorem 6 we have

$$\begin{aligned}
\vdash C_m \cdot C_m &= B_{m-1}(C_1 \cdot C_1) \\
&= B_{m-1}(B_2 I) \quad \text{(Ax. } (CC)_1), \\
&= I \quad \text{(II B 1, Thm. 5; II B 2, Thm. 1).}
\end{aligned}$$

Theorem 2. $\vdash C_m \cdot C_{m+1} \cdot C_m = C_{m+1} \cdot C_m \cdot C_{m+1} \quad (m = 1, 2, \cdots)$.

Proof.

$$\begin{aligned}
\vdash C_m \cdot C_{m+1} \cdot C_m & \\
&= B_{m-1}(C_1 \cdot C_2 \cdot C_1) \quad \text{(II B 4, Thm. 3 and 6),} \\
&= B_{m-1}(C_2 \cdot C_1 \cdot C_2) \quad \text{(Ax. } (CC)_2), \\
&= C_{m+1} \cdot C_m \cdot C_{m+1}.
\end{aligned}$$

Theorem 3. $\vdash C_m \cdot C_{m+j} = C_{m+j} \cdot C_m$ *if* $j > 1$, $(m = 1, 2, \cdots)$.

Proof. Follows by §2, Theorem 5, when we put C_j instead of Y in equation a).

Theorem 4. $\vdash C_m \cdot W = W \cdot C_{m+1} \quad (m = 2, 3, 4, \cdots)$.

Proof. Follows similarly from §2, Theorem 5b.

Theorem 5. $\vdash W_m \cdot W_n = W_n \cdot W_{m+1} \quad (m \geq n = 1, 2, 3, \cdots)$.

Proof.

For $m = n$,

$$\begin{aligned}
\vdash W_m \cdot W_m & \\
&= B_{m-1}(W_1 \cdot W_1) \quad \text{(II B 3, Def. 2; II B 4, Thm. 6),} \\
&= B_{m-1}(W_1 \cdot W_2) \quad \text{(Ax. } (WW)), \\
&= W_m \cdot W_{m+1} \quad \text{(II B 3, Def. 2; II B 4, Thm. 6).}
\end{aligned}$$

For $m > n$, the theorem follows by §2, Theorem 5b.

SATZ 6. $W_{m+1} = C_m \cdot W_m \cdot C_{m+1} \cdot C_m \qquad (m = 1, 2, 3, \cdots)$.

Beweis: $\vdash C_m \cdot W_m$
$$\begin{aligned}
&= B_{m-1}(C_1 \cdot W_1) && \text{(II B 3, Def. 2; II B 4, Satz 6)},\\
&= B_{m-1}(W_2 \cdot C_1 \cdot C_2) && \text{(Ax. }(CW)),\\
&= W_{m+1} \cdot C_m \cdot C_{m+1} && \text{(II B 3, Def. 2; II B 4, Satz 6)}.
\end{aligned}$$
also $\begin{aligned}[t] W_{m+1} &= (W_{m+1} \cdot C_m \cdot C_{m+1}) \cdot C_{m+1} \cdot C_m && \text{(Satz 1)},\\
&= C_m \cdot W_m \cdot C_{m+1} \cdot C_m && \text{w.z.b.w.}
\end{aligned}$

SATZ 7. *Wenn* $\mathfrak{M}_1$ *und* $\mathfrak{M}_2$ *derselben Folge lauter Variablen entsprechen, dann* $\vdash \mathfrak{M}_1 = \mathfrak{M}_2$.

Beweis: In meiner oben zitierten Abhandlung gegeben. Die Voraussetzungen jenes Beweises sind in der Tat schon hier bewiesen, wie folgt:

dort		hier
Axiomschema	I	Satz 1
"	II	Satz 2
"	III	Satz 3
"	IV	Satz 4
"	V	Ax. (WC)
"	VI	Folgen aus Satz 5 durch Umkehrung des Beweises der
"	VII	Gleichungen (6) und (7) meiner zitierten Abhandlung.
Definition von	W_k	Satz 6.

Jener Beweis lässt sich aber vermöge der hier vorliegenden Entwicklungen bedeutend abkürzen. In der Tat können wir aus §2, Satz 5b und Ax. (CW) in einer den Beweisen von §3, Sätzen 2 und 4 ähnlicher Weise schliessen, dass ein $\mathfrak{M}$ in die Form $\mathfrak{W} \cdot \mathfrak{C}$ umgeformt werden kann, und dann weiter, wie im §3, Satz 1 nachweisen, dass $\mathfrak{W}$ sich in die Normalform umformen lässt. Dabei werden Lemmas 1 und 2 jener Abhandlung bewiesen. Für Lemmas 3 und 4 sind alternative Beweise schon in II C4, Sätzen 4 und 5 geliefert.

SATZ 8. *Jedes* $\mathfrak{M}$ *lässt sich in die Normalform umformen.*

Beweis: Dies ist im Laufe des Beweises von Satz 7 dargetan. (Lemmas 1 und 2 meiner früheren Abhandlung).–Der Satz folgt auch direkt aus Satz 7, §6 (unten) Satz 2, und II C 5, Satz 2.

§6. Zusammenfassung und Schluss.

SATZ 1. *Jeder reguläre Kombinator kann in die Normalform umgeformt werden.*

Theorem 6. $W_{m+1} = C_m \cdot W_m \cdot C_{m+1} \cdot C_m$ $(m = 1, 2, 3, \cdots)$.

Proof.

$$\begin{aligned}
\vdash C_m \cdot W_m &\\
= B_{m-1}(C_1 \cdot W_1) &\quad \text{(II B 3, Def. 2; II B 4, Thm. 6)},\\
= B_{m-1}(W_2 \cdot C_1 \cdot C_2) &\quad \text{(Ax. } (CW)),\\
= W_{m+1} \cdot C_m \cdot C_{m+1} &\quad \text{(II B 3, Def. 2; II B 4, Thm. 6)}.
\end{aligned}$$

Hence

$$\begin{aligned}
W_{m+1} &= (W_{m+1} \cdot C_m \cdot C_{m+1}) \cdot C_{m+1} \cdot C_m \quad \text{(Thm. 1)},\\
&= C_m \cdot W_m \cdot C_{m+1} \cdot C_m \quad \text{qed.}
\end{aligned}$$

Theorem 7. *If $\mathfrak{M}_1$ and $\mathfrak{M}_2$ conform to the same sequence of variables only then $\vdash \mathfrak{M}_1 = \mathfrak{M}_2$.*

Proof. Given in my above cited paper. The hypotheses for that proof are in fact already proven here as follows:

	there		here
	Axiomschema	I	Theorem 1
	"	II	Theorem 2
	"	III	Theorem 3
	"	IV	Theorem 4
	"	V	Ax. (WC)
	"	VI	Both VI and VII follow from Thm. 5
	"	VII	by a reversal of the proofs of equations
			(6) and (7) in my above cited paper.
	Definition of	W_k	Theorem 6.

By virtue of the present developments here, that proof can be substantially shortened. In fact, by §2, Theorem 5b and Ax. (CW), we can derive in a fashion similar to the proofs of §3, Theorems 2 and 4, that an $\mathfrak{M}$ can be converted into the form $\mathfrak{W} \cdot \mathfrak{C}$ and then further deduce, as in §3, Theorem 1, that $\mathfrak{W}$ can be converted into normal form. This way Lemmas 1 and 2 of that paper are proven. For Lemmas 3 and 4 alternative proofs are already given in II C4, Theorems 4 and 5.

Theorem 8. *Every $\mathfrak{M}$ can be converted into normal form.*

Proof. This was done during the proof of Theorem 7. (Lemmas 1 and 2 of my earlier paper). –The theorem also follows directly from Theorem 7, §6 (below) Theorem 2, and II C5, Theorem 2.

§6. *Summary and conclusion.*

Theorem 1. *Every regular combinator can be converted into normal form.*

Beweis: Jeder reguläre Kombinator X lässt sich in die Form $(\Omega \cdot \mathfrak{B})$, wo $\mathfrak{B}$ in der Normalform steht, umformen (§3, Sätze 1 und 6). Dieses Ω lässt sich in die Form $(\mathfrak{K} \cdot \mathfrak{M})$ umformen, wo $\mathfrak{K}$ in der Normalform ist (§4, Sätze 1 und 2). Endlich lässt sich $\mathfrak{M}$ in die Normalform umformen (§4, Satz 8). Also kann X in die Normalform $(\mathfrak{R} \cdot \mathfrak{W} \cdot \mathfrak{C} \cdot \mathfrak{B})$ umgeformt werden.

SATZ 2. *Jeder reguläre Kombinator entspricht einer normalen Folge lauter Variablen, und zwar im ersten Sinne.*

Beweis: Die einzelnen Glieder eines regulären Kombinator entsprechen solchen Folgen (II B 3; II C 3, Satz 1; II B 2). Daher entspricht das Produkt einer solchen Folge (II C 2, Satz 2). Dass er der Folge im ersten Sinne entspricht, ist aus dem Beweis von II C2, Satz 2 ohne weiteres ersichtlich.

SATZ 3. *Wenn X_1 und X_2 reguläre Kombinatoren sind, wofür $\vdash X_1 = X_2$; dann sind X_1 und X_2 äquivalent in dem dritten Sinne.*

Beweis: Nach II C 1, Satz 11, und Satz 2 sind X_1 und X_2 im ersten Sinne äquivalent also entsprechen sie beide einer gemeinsamen Folge lauter Variablen. Nach Satz 2 entsprechen sie dieser Folge im ersten Sinne. Also ist der Sinn der Äquivalenz zwischen X_1 und X_2 der dritte.

SATZ 4. *Wenn X_1 und X_2 reguläre, derselben Folge von lauter Variablen entsprechende Kombinatoren sind; dann $\vdash X_1 = X_2$.*

Beweis: Sind Y_1 und Y_2 reguläre, in der Normalform stehende Kombinatoren, in welche X_1 bzw. umgeformt werden können (Satz 1), so entsprechen Y_1 und Y_2 derselben Folge wie X_1 und X_2 (Hp. und Satz 3).

Sind $\quad Y_1 \equiv \mathfrak{K}_1 \cdot \mathfrak{M}_1 \cdot \mathfrak{B}_1$ und $Y_2 \equiv \mathfrak{K}_2 \cdot \mathfrak{M}_2 \cdot \mathfrak{B}_2$,
dann $\quad\quad \vdash \mathfrak{B}_1 = \mathfrak{B}$ und $\vdash \mathfrak{K}_1 = \mathfrak{K}_2 \quad$ (II C 5, Satz 2),
und $\quad\quad\quad \vdash \mathfrak{M}_1 = \mathfrak{M}_2 \quad\quad\quad\quad\quad$ (§5, Satz 7).
Daher $\quad\quad \vdash Y_1 = Y_2$.
also $\quad\quad\quad \vdash X_1 = X_2, \quad\quad\quad\quad\quad$ w.z.b.w.

SATZ 5. *Damit für zwei reguläre Kombinatoren X_1 und $X_2 \vdash X_1 = X_2$ gilt, ist es notwendig und hinreichend, dass X_1 und X_2 im dritten Sinne äquivatent sind.*

Beweis: Klar aus Sätzen 3 und 4.

Festsetzung 1. Eine normale Folge ξ hat die Ordnung n wenn 1) es

Proof. Every regular combinator can be converted into the form $(\Omega \cdot \mathfrak{B})$, where $\mathfrak{B}$ is in normal form (§3, Theorems 1 and 6). This Ω can be converted into the form $(\mathfrak{K} \cdot \mathfrak{M})$ where $\mathfrak{K}$ is in normal form (§4, Theorems 1 and 2). Finally $\mathfrak{M}$ can be converted into normal form (§5, Theorem 8). Thus X can be converted into the normal form $(\mathfrak{K} \cdot \mathfrak{M} \cdot \mathfrak{C} \cdot \mathfrak{B})$.

Theorem 2. *Every regular combinator conforms to a normal sequence of variables only, and this in the first sense.*

Proof. The individual limbs of a regular combinator conform to such sequences (II B 3; II C 3, Theorem 1; II B 2). Hence the product conforms to such a sequence (II C2, Theorem 2). That it conforms to the sequence in the first sense is readily apparent by the proof of II C2, Theorem 2.

Theorem 3. *If X_1 and X_2 are regular combinators, for which $\vdash X_1 = X_2$, then X_1 and X_2 are equivalent in the third sense.*

Proof. By II C 1, Theorem 11, and Theorem 2, X_1 and X_2 are equivalent in the first sense, so they both conform to a common sequence of variables only. By Theorem 2, they conform to this sequence in the first sense. Thus the sense of the equivalence between X_1 and X_2 is the third.

Theorem 4. *If X_1 and X_2 are regular and conform to the same sequence of variables only, then $\vdash X_1 = X_2$.*

Proof. Let Y_1 and Y_2 be regular combinators in normal form, to which X_1 resp. X_2 can be converted (Theorem 1), then Y_1 and Y_2 conform to the same sequence as X_1 and X_2 (hypothesis and Theorem 3).

Let $Y_1 \equiv \mathfrak{K}_1 \cdot \mathfrak{M}_1 \cdot \mathfrak{B}_1$ and $Y_2 \equiv \mathfrak{K}_2 \cdot \mathfrak{M}_2 \cdot \mathfrak{B}_2$,
then $\vdash \mathfrak{B}_1 = \mathfrak{B}_2$ and $\vdash \mathfrak{K}_1 = \mathfrak{K}_2$ (II C 5, Thm. 2),
and $\vdash \mathfrak{M}_1 = \mathfrak{M}_2$ (§5, Thm. 7).
Hence $\vdash Y_1 = Y_2$.
Thus $\vdash X_1 = X_2$, qed.

Theorem 5. *For $\vdash X_1 = X_2$ to hold for two regular combinators X_1 and X_2, it is necessary and sufficient that X_1 and X_2 be equivalent in the third sense.*

Proof. Obvious by Theorems 3 and 4.

Convention 1. A normal sequence ξ has *order n* when 1) there

eine Kombination X von $x_0, x_1, x_2, \cdots, x_n$ gibt, sodass die Folge durch (Xx_{n+1}) bestimmt ist, und 2) n die keinste Zahl ist, wofür ein solches X existiert.

Es folgt aus dieser Festsetzung, dass jeder Kombinator der dem ξ entspricht, ihm mindestens mit der Ordnung $n+1$ entspricht.*

SATZ 6. *ξ sei eine normale Folge der Ordnung n, und X sei ein der Folge ξ entsprechender normaler Kombinator. Dann entspricht X der Folge ξ mit der Ordnung $n+1$.*

Beweis: Wir nehmen ein m so gross, dass X', wobei

$$X' \equiv Xx_0x_1x_2 \cdots x_m,$$

sich auf einen Abschnitt von ξ reduziert. Wenn in dieser Reduktion keine der Variablen $x_{n+1}, x_{n+2}, \cdots, x_m$, gestört werden, so ist der Satz bewiesen. Sonst führen wir die Reduktion von X' ohne Störung von $x_{n+1}, x_{n+2}, \cdots, x_m$ soweit fort, bis wir auf einen Ausdruck der Form

$$Y(Zx_0)y_1y_2 \cdots y_q$$

kommen (wo Y ein Glied von X_k, Z† ein Produkt solcher Glieder ist, und $y_1 \cdots y_q$ Kombinationen von $x_1x_2 \cdots x_m$ sind), sodass eine weitere Reduktion auf einen Ausdruck derselben Form ohne Störung von $x_{n+1} \cdots x_m$ nicht möglich ist. Wir unterscheiden dann vier Fälle:

1) Y ist ein K_p. Dann wird ein $x_s, s > n$, in der weiteren Reduktion ausgelassen. Weil durch Reduktionprozesse keine Variablen eingesetzt werden, so bleibt x_s ausgelassen bis zur Ende der Reduktion von X'. Weil dieses x_s nicht in ξ ausgelassen ist, kann X nicht der Folge ξ entsprechen.

2) Y ist ein W_p. Dann wird in der weiteren Reduktion ein x_s, $s > n$, verdoppelt. Weil X normal ist, so kann kein Glied der Form K_p in Z vorkommen; also bleibt x_s verdoppelt bis zur Ende. Weil x_s nicht in ξ verdoppelt ist, so kann X auch in diesem Falle nicht der Folge ξ entsprechen.

3) Y ist ein C_p. In diesem Falle führen wir die Reduktion fort, bis wir an einen Ausdruck der obigen Form ankommen, wo nun Y das C_p mit höchstem Index ist. Durch dieses C_p wird ein höchstes x_s $s > n$ mit einer niedrigeren x_t vertauscht, und weil dieses C_p nur einmal vorkommt (§1, Fest-

---

*Wir haben hier $n+1$, nicht n, weil ich die Variable x_0 zugelassen habe. Die Behauptung folgt, weil in jeder Reduktion auf einen Abschnitt von ξ die Variable x_n gestört werde muss.

†Streng genommen, können wir statt (Zx_0) einen Ausdruck haben, worauf (Zx_0) sich rednziert; aber dies stört den Kern des Beweises gar nicht.

is a combination X of $x_0, x_1, x_2, \cdots,, x_n$ such that this sequence is determined by (Xx_{n+1}), and 2) n is the smallest number for which such an X exists.

It follows by this convention that every combinator which conforms to the sequence ξ conforms to it at least with order $n+1$.*

Theorem 6. *Let ξ be a normal sequence of order n, and X be a normal combinator which conforms to ξ. Then X conforms to the sequence ξ with order $n+1$.*

Proof. We take an m so large that X', where

$$X' \equiv Xx_0x_1\cdots x_m,$$

reduces to a section of ξ. If in this reduction none of the variables $x_{n+1}, x_{n+2}, \cdots, x_m$ are disturbed, then the theorem is proven. Otherwise, we carry out the reduction of X' as far as possible without disturbing $x_{n+1}, x_{n+2}, \cdots, x_m$ until we reach an expression of the form

$$Y(Zx_0)y_1y_2\cdots y_q$$

(where Y is a limb of X_k, $Z^\dagger$ is a product of such limbs, and $y_1\cdots y_q$ are combinations of $x_1x_2\cdots x_m$) so that a further reduction to an expression of the same form without disturbance of $x_{n+1}\cdots x_m$ is not possible. Then we distinguish four cases:

1) Y is a K_p. Then an x_s, $s > n$, will be lost in the further reduction. Because during the reduction process no variables are replaced, then x_s remains lost until the end of the reduction of X'. Since this x_s is not lost in ξ, X cannot conform to the sequence ξ.

2) Y is a W_p. Then in the further reduction an x_s, $s > n$, will be doubled. Because X is normal then no limb of the form K_p can occur in Z; hence x_s remains doubled till the end. Since x_s is not doubled in ξ, so also in this case, X cannot conform to the sequence ξ.

3) Y is a C_p. In this case we carry out the reduction until we reach an expression of the above form, where now Y is the C_p with the highest index. By this C_p, the highest x_s, $s > n$, will be exchanged with a lower x_t, and because this C_p occurs only once (§1, convention

---

*We have $n+1$ here, not m, because I have admitted the variable x_0. The assertion follows because in every reduction to a section of ξ the variable x_n will be disturbed.

†Strictly speaking, we can have instead of (Zx_0) an expression to which (Zx_0) reduces; but this does not affect the core of the proof at all.

setzung 2), so kann x_s nie seine Stelle wieder erreichen. Aber dies widerspricht noch einmal der Voraussetzung, dass X der Folge ξ entspricht.

4) Y ist ein $B_p B_q$. Dann reduziert sich X' auf eine Kombination, worin mindestens ein x_s, $s > n_3$ eingeklammert ist. Daher entspricht X nicht der Folge ξ.

Diese vier Fälle erschöpfen alle möglichkeiten, weil Glieder der Form $B_m I$ in einem normalen Kombinator nicht vorkommen.

SATZ 7. *Wenn X eine beliebige normale Kombination von lauter Variablen ist, so gibt es einen normalen Kombinator, der sie darstellt.*

Beweis: Wir nehmen an, dass X eine normale Kombination der Variablen $x_0, x_1, \cdots, x_n$ ist. Y sei der normale Kombinator, welcher der durch X bestimmten Folge entspricht (II C 5, Satz 2). Die Ordnung dieses Entsprechens ist $\leq n + 1$ (Satz 6, Festselzung 1). Also muss $(Y x_0 x_1 x_2 \cdots x_n)$ sich aus X reduzieren, und daher wird ipso facto X durch Y dargestellt.

E. EIGENTLICHE KOMBINATOREN

§1. *Vorläufige Festsetzungen und Sätze.*

Festsetzung 1. Ein Kombinator heisst *eigentlich*, wenn er einer Folge lauter Variablen entspricht.

In diesem Abschnitte beweise ich, dass jeder eigentliche Kombinator in der Form $\mathfrak{R}I$, wo $\mathfrak{R}$ regulär ist, umgeformt werden kann. Daraus folgt, hinsichtlich der Ergebnisse des letzten Abschnitts, dass zwei derselben Folge entsprechende Kombinatoren immer gleich sind. Der Beweis der in Abschnitt A erwähnten Hauptsätze II und III wird hier vollzogen (der letzte für eigentliche Kombinatoren).

Festsetzung 2. Ausser den Gattungszeichen von II D 1, Festsetzung 4 benutze ich den Buchstaben $\mathfrak{R}$ für einen regulären Kombinator.

Festsetzung 3. Ein Kombinator heist *regulierbar*, wenn er in einen regulären Kombinator umgeformt werden kann; d. h. wenn es einen regulären Kombinator gibt, der ihm gleich ist.

SATZ 1. *Sind die Kombinatoren X und Y regulierbar, so ist auch $(X \cdot Y)$ regulierbar.*

Beweis: Nach den Voraussetzungen gibt es $\mathfrak{R}_1$ und $\mathfrak{R}_2$, sodass $\vdash X = \mathfrak{R}_1$ und $\vdash Y = \mathfrak{R}_2$, also $\vdash X \cdot Y = \mathfrak{R}_1 \cdot \mathfrak{R}_2$. $(\mathfrak{R}_1 \cdot \mathfrak{R}_2)$ ist aber regulär (dies folgt direkt aus II D 1, Festsetzung 1).

2), then x_s can never reach its place again. But this once again contradicts the hypothesis that X conforms to the sequence ξ.

4) Y is a $B_p B_q$. Then X' reduces to a combination, in which at least one x_s, $s > n$, is between parenthesis. Hence X does not conform to the sequence ξ.

These four cases are exhaustive because limbs of the form $B_m I$ do not occur in a normal combinator.

Theorem 7. *If X is an arbitrary normal combination of variables only, then there is a normal combinator that represents it.*

Proof. We assume that X is a normal combination of the variables $x_0, x_1, \cdots, x_n$. Let Y be the normal combinator which conforms to the sequence determined by X (II C 5, Theorem 2). The order of this conformity is $\leqq n + 1$ (Theorem 6, Convention 1). Thus $(Y x_0 x_1 x_2 \cdots x_n)$ reduces to X and hence by that very fact, X is represented by Y.

E. Proper Combinators

§1. *Preliminary conventions and theorems.*

Convention 1. A combinator is called *proper* when it conforms to a sequence of variables only.

In this section I will prove that every proper combinator can be converted into the form $\mathfrak{R}I$ where $\mathfrak{R}$ is regular. It follows with respect to the results of the last sections, that two combinators which conform to the same sequence are always equal. The proof of the main theorems II and III mentioned in Section A will be completed here (the latter for proper combinators).

Convention 2. Besides the generic signs of II D 1, Convention 4, I use the letter $\mathfrak{R}$ for a regular combinator.

Convention 3. A combinator is called *regularizing (regulierbar)* if it can be converted into a regular combinator; i.e., if there is a regular combinator which is equal to it.

Theorem 1. *If the combinators X and Y are regularizing, then $(X \cdot Y)$ is also regularizing.*

Proof. By the hypotheses there exist $\mathfrak{R}_1$ and $\mathfrak{R}_2$ such that $\vdash X = \mathfrak{R}_1$ and $\vdash Y = \mathfrak{R}_2$, hence $\vdash X \cdot Y = \mathfrak{R}_1 \cdot \mathfrak{R}_2$. But $(\mathfrak{R}_1 \cdot \mathfrak{R}_2)$ is regular (this follows directly from II D1, Convention 1).

SATZ 2. *Ist der Kombinator X regulierbar, so ist jedes $(\mathfrak{B}X)$ regulierbar.*

Beweis: Wenn $\mathfrak{B} \equiv I$ ist, klar.

Zunächst sei $\mathfrak{B} \equiv B_m$. Setzen wir dann

$$\vdash X = \mathfrak{R}, \qquad \mathfrak{R} \equiv X_1 \cdot X_2 \cdots X_n.$$

Dann $\vdash B_m X = B_m X_1 \cdot B_m X_2 \cdots B_m X_n$ (II B 4, Sätze 3 u. 6), und die rechte Seite ist regulär.

Es sei nun ein allgemeines $\mathfrak{B}$ gegeben. Wir können annehmen, dass $\mathfrak{B}$ in der Normalform steht. Dann folgt wenn $m_1 = 0$ ist (wo m_1 wie in II C 3, Satz 3 zu verstehen ist),

$$\vdash \mathfrak{B} = B\mathfrak{B}' \cdot B_{n_1}$$

also $\vdash \mathfrak{B}X = B\mathfrak{B}'(B_{n_1}X) = \mathfrak{B}' \cdot B_{n_1}X.$

Die rechte Seite ist regulierbar nach dem eben Bewiesenen und Satz 1 Dagegen sei $m_1 > 0$. Dann

$$\begin{aligned}\vdash \mathfrak{B}X &= B_{m_1}\mathfrak{B}'X = B(B_{m_1-1}\mathfrak{B}')X \\ &= B_{m_1-1}\mathfrak{B}' \cdot X.\end{aligned}$$

Die rechte Seite ist wieder regulierbar nach dem oben Gesagten und Satz 1.

SATZ 3. *Wenn X und Y beliebige Etwase sind, dann $\vdash XY = (X \cdot BY)I$.*

Beweis: Klar aus II B2, Satz 4, und II B 4, Satz 1.

SATZ 4. *Jeder Kombinator der Form $(\mathfrak{R}I)$ entspricht einer Folge lauter Variablen, und zwar in dem ersten Sinne.*

Beweis: n sei so gewählt, dass der Ausdruck $(\mathfrak{R}x_0 x_1 x_2 \cdot \cdot x_n)$ sich auf eine normale Kombination von $x_0, x_1, x_2, \cdots, x_n$ etwa $(x_0 y_1 y_2 \cdots y_q)$ ohne Auslassung von x_n reduziert (möglich nach II D6, Satz 2). Dann wird $(\mathfrak{R}I x_1 x_2 \cdots x_m)$ auf $(Iy_1y_2 \cdots y_q)$ im ersten Sinne reduziert. Dass sich die weitere Reduktion auf $(y_1 y_2 \cdot y_q)$ im ersten Sinne vollzieht, ist selbstverständlich. Also entspricht $(\mathfrak{R}I)$ der durch die eben geschilderte Kombination bestimmte Folge.

SATZ 5. *Eine notwendige und hinreichende Bedingung dafür, dass ein $(\mathfrak{R}I)$ einer normalen Folge entspricht, ist, dass es ein $\mathfrak{R}'$ und ein $\mathfrak{B}$ gibt, sodass*

$$\vdash \mathfrak{R} = B\mathfrak{R}' \cdot \mathfrak{B}.$$

Beweis: Die Bedingung ist hinreichend; denn ist sie erfüllt, so gilt

$$\vdash \mathfrak{R}I = (B\mathfrak{R}' \cdot \mathfrak{B})I = \mathfrak{R}' \cdot \mathfrak{B}I.$$

Theorem 2. *If the combinator X is regularizing, then every $(\mathfrak{B}X)$ is regularizing.*

Proof. If $\mathfrak{B} \equiv I$, obvious.

Now let $\mathfrak{B} \equiv B_m$. Then assume

$$\vdash X = \mathfrak{R}, \qquad \mathfrak{R} \equiv X_1 \cdot X_2 \cdots X_n.$$

Then $\vdash B_m X = B_m X_1 \cdot B_m X_2 \cdots B_m X_m$ (II B 4, Thm. 3 and 6), and the right hand side is regular.

Consider now a general $\mathfrak{B}$. We can assume that $\mathfrak{B}$ is in normal form. If $m_1 = 0$ (where m_1 is to be understood as in II C 3, Theorem 3), then it follows that

$$\vdash \mathfrak{B} = B\mathfrak{B}' \cdot B_{n_1}$$

hence $\qquad \vdash \mathfrak{B}X = B\mathfrak{B}'(B_{n_1}X) = \mathfrak{B}' \cdot B_{n_1}X.$

This right hand side is regularizing by what has just been proved and by Theorem 1.

On the other hand, assume $m_1 > 0$. Then

$$\vdash \mathfrak{B}X = B_{m_1}\mathfrak{B}'X = B(B_{m_1-1}\mathfrak{B}')X$$
$$= B_{m_1-1}\mathfrak{B}' \cdot X.$$

The right hand side is again regularizing by what was said above and Theorem 1.

Theorem 3. *If X and Y are arbitrary entities then $\vdash XY = (X \cdot BY)I$.*

Proof. Obvious by II B 2, Theorem 4, and II B 4, Theorem 1.

Theorem 4. *Every combinator of the form $(\mathfrak{R}I)$ conforms to a sequence of variables only, and this in the first sense.*

Proof. Let n be chosen such that the expression $(\mathfrak{R}x_0 x_1 x_2 \cdots x_n)$ reduces to a normal combination of variables $x_0, x_1, x_2, \cdots x_n$, like $(x_0 y_1 y_2 \cdots y_q)$ without loss of x_n (possible by II D 6, Theorem 2). Then $(\mathfrak{R}I x_1 x_2 \cdots x_n)$ reduces to $(I y_1 y_2 \cdots y_q)$ in the first sense. That the further reduction to $(y_1 y_2 \cdots y_q)$ takes place in the first sense is self-evident. Hence $(\mathfrak{R}I)$ conforms to the sequence determined by the combination just described.

Theorem 5. *A necessary and sufficient condition for an $(\mathfrak{R}I)$ to conform to a sequence is that there exist an $\mathfrak{R}'$ and a $\mathfrak{B}$ such that*

$$\vdash \mathfrak{R} = B\mathfrak{R}' \cdot \mathfrak{B}$$

Proof. The condition is sufficient; because if it is satisfied, then

$$\vdash \mathfrak{R}I = (B\mathfrak{R}' \cdot \mathfrak{B})I = \mathfrak{R}' \cdot \mathfrak{B}I.$$

($\mathfrak{B}I$) ist regulierbar nach Satz 2; also ist ($\mathfrak{R}I$) regulierbar nach Satz 1. Daher entspricht ($\mathfrak{R}I$) einer normalen Folge (Satz 4; II D 6, Satz 2; II C 1, Satz 11).

Die Bedingung ist notwendig. In der Tat sei angenommen, dass ($\mathfrak{R}Ix_1x_2\cdots x_n$) sich auf eine normale Kombination V von $x_1, x_2, \cdots, x_n$ reduziert. Dann erscheint x_1 in V vereinzelt und an der ersten Stelle. $\mathfrak{R}$ werde in die Normalform umgeformt, etwa

$$\vdash \mathfrak{R} = \mathfrak{K} \cdot \mathfrak{W} \cdot \mathfrak{C} \cdot \mathfrak{B}.$$

Dann ist $\mathfrak{K}$ von der Faktor K_1 frei, weil sonst x_1 in V ausfallen würde, also

$$\vdash \mathfrak{K} = B\mathfrak{K}'.$$

Gleichfalls ist $\mathfrak{W}$ von der Faktor W_1 frei, weil sonst x_1 in V verdoppelt sein würde, also $\vdash \mathfrak{W} = B\mathfrak{W}'$. Weiter entspricht $\mathfrak{C}$ einer durch eine Permutation der Variablen $x_2, x_3, \cdots, x_m$ bestimmten Folge, also ist $\mathfrak{C}$ in ein Produkt von $C_2, C_3, \cdots, C_{n-1}$, umformbar* und daher $\vdash \mathfrak{C} = B\mathfrak{C}'$. Aus den letzten drei Formeln folgt

$$\vdash \mathfrak{R} = B(\mathfrak{K}' \cdot \mathfrak{W}' \cdot \mathfrak{C}') \cdot \mathfrak{B} \qquad \text{w.z.b.w.}$$

SATZ 6. *Zu jeder Folge lauter Variablen gibt es ein $\mathfrak{R}_1$, und zwar ein normales $\mathfrak{R}_1$ ohne Glieder der Form B_n, sodass ($\mathfrak{R}_1 I$) der Folge entspricht. Gibt es überdies ein anderes der Folge entsprechendes $\mathfrak{R}_2$, so gilt für ein durch $\mathfrak{R}_2$ bestimmtes n*

$$\vdash \mathfrak{R}_2 = \mathfrak{R}_1 \cdot B_n.$$

Beweis: Wir nehmen an, die Variablen in der gegebenen Folge sind $x_1, x_2, x_3, \cdots$. Die Folge sei etwa

(1) $\qquad x_j y_1 y_2 y_3 \cdots \qquad j \geq 1.$

wo y_i eine Kombination gewissen x's ist. $\mathfrak{R}_1$ sei ein normaler Kombinator, welcher der Folge

(2) $\qquad x_0 x_j y_1 y_2 y_3 \cdots$

entspricht. Dann entspricht ($\mathfrak{R}_1 I$) der gegebenen Folge nach dem Beweis von Satz 4. Enthält $\mathfrak{R}_1$ ein Glied der Form B_n, so müsste $\mathfrak{R}_1$, weil es normal ist, von der Form ($\mathfrak{R}_1' \cdot B_n$) sein; aber in diesem Falle würde $\mathfrak{R}_1$ einer Folge entsprechen, worin eine Anfangsklammer links von der zweiten Variablen steht. Weil (2) diese Form nicht hat, so erfüllt $\mathfrak{R}_1$ die Bedingungen des ersten Teils des Satzes.

---

*Vgl. Beweis von II C 4, Satz 5.

($\mathfrak{B}I$) is regularizing by Theorem 2; thus ($\mathfrak{R}I$) is regularizing by Theorem 1. Therefore ($\mathfrak{R}I$) conforms to a normal sequence (Theorem 4; II D 6, Theorem 2; II C 1, Theorem 11).

This condition is necessary. In fact assume that ($\mathfrak{R}Ix_1x_2\cdots x_n$) reduces to a normal combination V of $x_1, x_2, \cdots x_n$. Then x_1 appears isolated in V and in the first place. $\mathfrak{R}$ will convert into this normal form, say

$$\vdash \mathfrak{R} = \mathfrak{K} \cdot \mathfrak{M} \cdot \mathfrak{C} \cdot \mathfrak{B}.$$

Then $\mathfrak{R}$ is free from the factor K_1, because otherwise x_1 in V is lost; hence

$$\vdash \mathfrak{K} = B\mathfrak{K}'.$$

Likewise $\mathfrak{B}$ is free from the factor W_1, because otherwise x_1 in V would have doubled; hence $\vdash \mathfrak{M} = B\mathfrak{M}'$. Furthermore, $\mathfrak{C}$ conforms to a sequence determined by a permutation of the variables $x_2, x_3, \cdots x_m$, hence $\mathfrak{C}$ can be converted* into a product of $C_2, C_3, \cdots, C_{n-1}$ and therefore $\vdash \mathfrak{C} = B\mathfrak{C}'$. By the last three formulas it follows that

$$\vdash \mathfrak{R} = B(\mathfrak{R}' \cdot \mathfrak{M}' \cdot \mathfrak{C}') \cdot \mathfrak{B} \quad \text{qed.}$$

Theorem 6. *To every sequence of variables only there is an $\mathfrak{R}_1$, which is normal and without limbs of the form B_n, such that ($\mathfrak{R}_1I$) conforms to the sequence. If there is another such $\mathfrak{R}_2$ for which ($\mathfrak{R}_2I$) conforms*[22] *to the sequence, then for an n determined by $\mathfrak{R}_2$*

$$\vdash \mathfrak{R}_2 = \mathfrak{R}_1 \cdot B_n.$$

Proof. We assume that the variables in the given sequence are $x_1, x_2, x_3, \cdots$ Let the sequence be

(1) $\qquad x_j y_1 y_2 y_3 \cdots \qquad j \geqq 1.$

where y_i is a combination of some x's. Let $\mathfrak{R}_1$ be a normal combinator which conforms to the sequence

(2) $\qquad x_0 x_j y_1 y_2 y_3 \cdots$

Then by the proof of Theorem 4, ($\mathfrak{R}_1I$) conforms to the given sequence. If $\mathfrak{R}_1$ contains a limb of the form B_n, then because it is normal, it must be of the form ($\mathfrak{R}'_1 \cdot B_n$); but in this case $\mathfrak{R}_1$ would conform to a sequence in which an opening parenthesis appears to the left of the second variable. Because (2) does not have this form, then $\mathfrak{R}_1$ satisfies the condition of the first part of the theorem.

---

*cf. proof of II C 4, Theorem 5.

Nun sei $\mathfrak{R}_2$ irgendein regulärer Kombinator derart, dass $(\mathfrak{R}_2 I)$ der gegebenen Folge entspricht. Wir können ohne Beschränkung der Allgemeinheit annehmen, dass $\mathfrak{R}_2$ normal ist (II D 6, Satz 1). Wenn $\mathfrak{R}_2$ Glieder der Form B_n enthält, so gibt es ein $\mathfrak{R}_2'$ ohne solche Glieder, und ein B_n, sodass
$$\vdash \mathfrak{R}_2 = \mathfrak{R}_2' \cdot B_n.$$
Dann gilt $\vdash \mathfrak{R}_2 I = \mathfrak{R}_2'(B_n I) = \mathfrak{R}_2' I$ \quad (II B 2, Satz 1; II B 2, Satz 1). Im entgegengesetzten Falle setzen wir $\mathfrak{R}_2' \equiv \mathfrak{R}_2 \cdot \mathfrak{R}_2'$ entspricht in den beiden Fällen einer Folge der Form
$$x_0 x_k z_1 z_2 z_3 \cdots,$$
(d. h. ohne Klammern vor der zweiten Variable.) Daher entspricht $\mathfrak{R}_2 I$ nach dem Beweis von Satz 4 der Folge:
$$x_k z_1 z_2 z_3 \cdots.$$
Weil dies mit der gegebenen Folge übereinstimmen muss, so ist $k = j$, $z_1 = y_1, z_2 = y_2$ u. s. w. $\mathfrak{R}_2'$ entspricht daher derselben Folge wie $\mathfrak{R}_1$. Also:
$$\vdash \mathfrak{R}_2' = \mathfrak{R}_1 \quad \text{(II D 6, Satz 4).}$$
$$\therefore \vdash \mathfrak{R}_2 = \mathfrak{R}_1 \cdot B_n \quad \text{w.z.b.w.}$$

Festseinung 4. Eine von der Variablen x_0 frei Folge ξ heisst der Ordnung n, wenn 1) es eine Kombination X von $x_1, x_2, \cdots x_n$ gibt, sodass die Folge durch $X x_{n+1}$ bestimmt wird, 2) n die kleinste Zahl dieser Beschaffenheit ist.

SATZ 7 *Dass $(\mathfrak{R}_1 I)$ von Satz 6 entspricht seiner Folge mit der Ordnung, die mit der Ordnung der Folge selbst übereinstimmt.*

Beweis: Das $\mathfrak{R}_1$ entspricht seiner normalen Folge mit der Ordnung $n + 1$, wo n die Ordnung der Folge selbst ist. (II D 6, Satz 6). Wie im Satz 4 folgt daraus, dass $(\mathfrak{R}_1 I)$ seiner Folge mit der Ordnung n entspricht.

SATZ 8. *Zu jeder Kombination lauter Variablen gibt es mindestens einen Kombinator, der sie darstellt.*

Beweis: Folgt aus Sätzen 6 und 7.

§2. *Die Kombinatoren Γ und eine Verallgemeinerung der kommutativen Gesetze.* Diese Sätze sind Hilfssätze für §3 unten.

Def. 1. \quad $\Gamma_1 \equiv C_1; \quad \Gamma_{n+1} \equiv \Gamma_n \cdot C_{n+1}, \quad (n = 1, 2, 3, \cdots).$

SATZ 1. $\vdash \Gamma_n = C_1 \cdot C_2 \cdots C_n.$

Beweis: Klar.

Now let $\mathfrak{R}_2$ be a regular combinator such that $(\mathfrak{R}_2 I)$ conforms to the given sequence. Without loss of generality we can assume that $\mathfrak{R}_2$ is normal (II D6, Theorem 1). If $\mathfrak{R}_2$ has limbs of the form B_n, then there is an $\mathfrak{R}_2'$ without such limbs, and a B_n such that

$$\vdash \mathfrak{R}_2' = \mathfrak{R}_2 \cdot B_n.$$

Hence $\vdash \mathfrak{R}_2 I = \mathfrak{R}_2'(B_n I) = \mathfrak{R}_2' I$ (II B 2, Thm. 1; II B 2, Thm. 1).

In the opposite case we assume $\mathfrak{R}_2' \equiv \mathfrak{R}_2 \cdot \mathfrak{R}_2'$ conforms in both cases to a sequence of the form

$$x_0 x_k z_1 z_2 z_3 \cdots,$$

(i.e., without parentheses before the second variable.) Hence $\mathfrak{R}_2 I$ conforms by the proof of theorem 4 to the sequence:

$$x_k z_1 z_2 z_3 \cdots.$$

Because this must coincide with the given sequence, then $k = j$, $z_1 = y_1$, $z_2 = y_2$ etc., hence $\mathfrak{R}_2'$ conforms to the same sequence as $\mathfrak{R}_1$. Thus:

$$\vdash \mathfrak{R}_2' = \mathfrak{R}_1 \quad \text{(II D 6, Theorem 4)}.$$
$$\therefore \vdash \mathfrak{R}_2 = \mathfrak{R}_1 \cdot B_n \quad \text{qed}.$$

Convention 4. A sequence ξ free of the variable x_0 is said to have *order* n if 1) there is a combination X of $x_1, x_2, \cdots, x_n$ such that the sequence is determined by $X x_{n+1}$, 2) n is the smallest number of this nature.

Theorem 7. *The $(\mathfrak{R}_1 I)$ of Theorem 6 conforms to its sequence with an order which coincides with the order of the sequence itself.*

Proof. The $\mathfrak{R}_1$ conforms to its normal sequence with the order $n+1$, where n is the order of the sequence itself. (II D6, Theorem 6). It follows as in Theorem 4 that $(\mathfrak{R}_1 I)$ conforms to its sequence with the order n.

Theorem 8. *To every combination of variables only there is at least one combinator that represents it.*

Proof. Follows by Theorems 6 and 7.

§2. *The combinators Γ and a generalisation of the commutative law.*

These theorems are lemmas for §3 below.

Def. 1. $\quad \Gamma_1 \equiv C_1; \quad \Gamma_{n+1} \equiv \Gamma_n \cdot C_{n+1}, \quad (n = 1, 2, 3, \cdots).$

Theorem 1. $\vdash \Gamma_n = C_1 \cdot C_2 \cdots \cdots C_n.$

Proof. Obvious.

SATZ 2. *Wenn $X_0, X_1, \cdots, X_n, Y$ beliebige Etwase sind, so gilt*
$\vdash \Gamma_n X_0 Y X_1 X_2 \cdots X_n = X_0 X_1 \cdots X_n Y.$

Beweis: Für $n = 1$, klar aus Regel C.

Ist nun der Satz für ein bestimmtes n angenommen, dann wird er für $n + 1$ wie folgt bewiesen:

$$\begin{aligned}
&\vdash \Gamma_{n+1} X_0 Y X_1 \cdots X_{n+1} \\
&= \Gamma_n (C_{n+1} X_0) X_1 Y X_2 \cdots X_n && \text{(Def. 1; II B 4, Satz 1)}, \\
&= C_{n+1} X_0 X_1 X_2 \cdots X_n Y X_{n+1} && \text{(Voraussetzung)}, \\
&= X_0 X_1 X_2 \cdots X_n X_{n+1} Y && \text{(II B 3, Satz 2)}.
\end{aligned}$$

Also folg der Satz durch Induktion.

SATZ 3. $\vdash \Gamma_{n+1} = C_1 \cdot B\Gamma_n.$

Beweis: Für $n = 1$ klar.

Ist der Satz für ein bestimmtes n angenommen, so gilt für dieses n

$$\begin{aligned}
&\vdash \Gamma_{n+2} \\
&= \Gamma_{n+1} \cdot C_{n+2} && \text{(Def. 1)}, \\
&= C_1 \cdot B\Gamma_n \cdot C_{n+2} && \text{(Hp.)}, \\
&= C_1 \cdot B(\Gamma_n \cdot C_{n+1}) && \text{(II B 3, Def. 1; II B 4, Sätze 2, 3)}, \\
&= C_1 \cdot B\Gamma_{n+1} && \text{(Def. 1)}.
\end{aligned}$$

Also folgt der Satz durch Induktion.

SATZ 4. $BB \cdot \Gamma_n = \Gamma_{n+1} \cdot B.$

Beweis: Für $n = 1$ ist dies in II D 1, Satz 2 bewiesen.

Ist der Satz für ein bestimmtes n angenommen, dann

$$\begin{aligned}
\vdash BB \cdot \Gamma_{n+1} &= BB \cdot \Gamma_n \cdot C_{n+1} && \text{(Def. 1)}, \\
&= \Gamma_{n+1} \cdot B \cdot C_{n+1} && \text{(Hp.)}, \\
&= \Gamma_{n+1} \cdot C_{n+2} \cdot B && \text{(II D 2, Satz 4)}, \\
&= \Gamma_{n+2} \cdot B && \text{(Def. 1)}.
\end{aligned}$$

Also wird der Satz durch Induktion bewiesen.

SATZ 5. *Wenn Y ein beliebiges Etwas ist; dann*

$$\vdash B_p(C_1 B_{n+1} Y) = \Gamma_{p+1}(B_p B_{n+1})Y, \qquad (p = 0, 1, 2, 3, \cdots).$$

Beweis: Definieren wir vorübergehend

$$X_p \equiv \Gamma_p(B_{p-1} B_{n+1})Y, \qquad (p = 1, 2, 3, \cdots),$$

dann $\qquad X_1 \equiv C_1 B_{n+1} Y,$

Theorem 2. *If $X_0, X_1, \cdots, X_n, Y$ are arbitrary entities then*
$\vdash \Gamma_n X_0 Y X_1 X_2 \cdots X_n = X_0 X_1 \cdots X_n Y.$

Proof. For $n = 1$, obvious by rule C.

Assume now the theorem holds for some n, then it will be proven for $n + 1$ as follows:

$\vdash \Gamma_{n+1} X_0 Y X_1 \cdots X_{n+1}$
$\quad = \Gamma_n (C_{n+1} X_0) Y X_1 X_2 \cdots X_{n+1}$ (Def. 1; II B 4, Thm. 1),
$\quad = C_{n+1} X_0 X_1 X_2 \cdots X_n Y X_{n+1}$ (hypothesis),
$\quad = X_0 X_1 X_2 \cdots X_n X_{n+1} Y$ (II B 3, Thm. 2).

Hence the theorem follows by induction.

Theorem 3. $\vdash \Gamma_{n+1} = C_1 \cdot B\Gamma_n.$

Proof. For $n = 1$, obvious.
Assume now the theorem holds for some n, then we have for this n

$\vdash \Gamma_{n+2}$
$\quad = \Gamma_{n+1} \cdot C_{n+2}$ (Def. 1),
$\quad = C_1 \cdot B\Gamma_n \cdot C_{n+2}$ (hypothesis),
$\quad = C_1 \cdot B(\Gamma_n \cdot C_{n+1})$ (II B 3, Def. 1; II B 4, Thm. 2, 3),
$\quad = C_1 \cdot B\Gamma_{n+1}$ (Def. 1).

Hence the theorem follows by induction.

Theorem 4. $BB \cdot \Gamma_n = \Gamma_{n+1} \cdot B.$

Proof. For $n = 1$, this is proven in II D 1, Theorem 2.
Assume now the theorem holds for some n, then

$\vdash BB \cdot \Gamma_{n+1} = BB \cdot \Gamma_n \cdot C_{n+1}$ (Def. 1),
$\quad = \Gamma_{n+1} \cdot B \cdot C_{n+1}$ (hypothesis),
$\quad = \Gamma_{n+1} \cdot C_{n+2} \cdot B$ (II D 2, Thm. 4),
$\quad = \Gamma_{n+2} \cdot B$ (Def. 1).

Thus the theorem is proven by induction.

Theorem 5. *If Y is an arbitrary entity; then*

$$\vdash B_p(C_1 B_{n+1} Y) = \Gamma_{p+1}(B_p B_{n+1}) Y, \qquad (p = 0, 1, 2, 3, \cdots).$$

Proof. If we define temporarily

$$X_p \equiv \Gamma_p(B_{p-1} B_{n+1}) Y, \qquad (p = 1, 2, 3, \cdots),$$

then $\quad X_1 = C_1 B_{n+1} Y,$

und

$$\begin{aligned}
\vdash BX_p &= B_2 B\Gamma_p(B_{p-1}B_{n+1})Y && \text{(II B 1, Satz 3)}, \\
&= (BB \cdot \Gamma_p)(B_{p-1}B_{n+1})Y && \text{(II B 1, Satz 5; II B 4, Def. 1)}, \\
&= \Gamma_{p+1}(B(B_{p-1}B_{n+1}))Y && \text{(Satz 4; II B 4, Satz 1)}, \\
&= \Gamma_{p+1}(B_p B_{n+1})X && \text{(II B 1, Satz 5)}, \\
&= X_{p+1}.
\end{aligned}$$

Also folgt der Satz aus II B 1, Satz 4.

SATZ 6. *Wenn X, Y beliebige Etwase sind, so gilt*

$$\Gamma_{p+1}(B_p B_{n+1} \cdot X)Y = \Gamma_{p+2}(B_{p+1}B_{n+1})YX.$$

Beweis:

$$\begin{aligned}
\vdash \Gamma_{p+1}&(B_p B_{n+1} X)Y \\
&= \Gamma_{p+1}(B_{p+1}B_{n+1}X)Y && \text{(II B 4, Def. 1; II B 1, Satz 5)}, \\
&= B\Gamma_{p+1}(B_{p+1}B_{n+1})XY && \text{(Reg. } B), \\
&= (C_1 \cdot B\Gamma_{p+1})(B_{p+1}B_{n+1})YX && \text{(II B 4, Satz 1; Reg. } C), \\
&= \Gamma_{p+2}(B_{p+1}B_{n+1})YX. && \text{(Satz 3)},
\end{aligned}$$

SATZ 7. *Wenn XY Etwase sind, und Y das Kommutativgesetz*

$$\vdash C_1 B_{m+1} Y = BY \cdot B_n$$

erfüllt; dann

$$\vdash \Gamma_{p+1}(B_p B_{n+1} \cdot X)Y = B_{p+1}Y \cdot B_p B_n \cdot X.$$

Beweis: Nach den Voraussetzungen,

$$\begin{aligned}
\vdash \Gamma_{p+1}&(B_p B_{m+1} \cdot X)Y \\
&= \Gamma_{p+2}(B_{p+1}B_{m+1})YX && \text{(Satz 6)}, \\
&= B_{p+1}(C_1 B_{m+1} Y)X && \text{(Satz 5)}, \\
&= B_p(C_1 B_{m+1} Y) \cdot X && \text{(II B 1, Satz 5; II B 4, Def. 1)}, \\
&= B_p(BY \cdot B_n) \cdot X && \text{(Hp.)}, \\
&= B_{p+1}Y \cdot B_p B_n \cdot X && \text{II B 4, Satz 6; II B 1, Satz 5)}.
\end{aligned}$$

§3. *Darstellung der allgerneinen Kombinationen.*

Festsetzung 1. Ein Ausdruck X der Form

$$(\mathfrak{R} I Y_1 Y_2 \cdots Y_p x_1 x_2 \cdots x_n),$$

wo die Y_i Etwase sind, *reduziert sich formal* auf einen Ausdruck Z, wenn mit Behandlung der Y_i als Variablen eine Reduktion von X auf Z sich durchführen lässt; oder, falls man es genauer haben will, wenn der Ausdruck $(\mathfrak{R} I x_1 x_2 \cdots x_{n+p})$ sich auf ein solches Z' reduziert, dass durch Einsetzung von Y_i statt x_i für $i = 1, 2, \cdots, p$, und von x_{i-p} statt x_i für $i = p+1, p+2, \cdots, p+n$ in Z' der Ausdruck Z erzielt wird.

and $\vdash BX_p$
$$\begin{align}
&= B_2 B \Gamma_p (B_{p-1} B_{n+1}) Y && \text{(II B 1, Thm. 3)},\\
&= (BB \cdot \Gamma_p)(B_{p-1} B_{n+1}) Y && \text{(II B 1, Thm. 5; II B 4, Def. 1)},\\
&= \Gamma_{p+1}(B(B_{p-1} B_{n+1})) Y && \text{(Thm. 4; II B 4, Thm. 1)},\\
&= \Gamma_{p+1}(B_p B_{n+1}) X && \text{(II B 1, Thm. 5)},\\
&= X_{p+1}.
\end{align}$$

Thus the theorem follows by II B1, Theorem 4.

Theorem 6. *If X and Y are arbitrary entities then*
$$\Gamma_{p+1}(B_p B_{n+1} \cdot X) Y = \Gamma_{p+2}(B_{p+1} B_{n+1}) Y X.$$

Proof.
$$\begin{align}
\vdash &\, \Gamma_{p+1}(B_p B_{n+1} \cdot X) Y \\
&= \Gamma_{p+1}(B_{p+1} B_{n+1} X) Y && \text{(II B 4, Def. 1; II B 1, Thm. 5)},\\
&= B \Gamma_{p+1}(B_{p+1} B_{n+1}) X Y && \text{(Rule B)},\\
&= (C_1 \cdot B\Gamma_{p+1})(B_{p+1} B_{n+1}) Y X && \text{(II B 4, Thm. 1; rule C)},\\
&= \Gamma_{p+2}(B_{p+1} B_{n+1}) Y X && \text{(Thm. 3)},
\end{align}$$

Theorem 7. *If X, Y are entities, and Y satisfies the commutative law*
$$\vdash C_1 B_{m+1} Y = BY \cdot B_n$$
then
$$\vdash \Gamma_{p+1}(B_p B_{m+1} \cdot X) Y = B_{p+1} Y \cdot B_p B_n \cdot X.$$

Proof. By the hypotheses,
$$\begin{align}
\vdash &\, \Gamma_{p+1}(B_p B_{m+1} \cdot X) Y \\
&= \Gamma_{p+2}(B_{p+1} B_{m+1}) Y X && \text{(Thm. 6)},\\
&= B_{p+1}(C_1 B_{m+1} Y) X && \text{(Thm. 5)},\\
&= B_p(C_1 B_{m+1} Y) \cdot X && \text{(II B 1, Thm. 5; II B 4, Def. 1)},\\
&= B_p(BY \cdot B_n) \cdot X && \text{(hypothesis)},\\
&= B_{p+1} Y \cdot B_p B_n \cdot X && \text{II B 4, Thm. 6; II B 1, Thm. 5)}.
\end{align}$$

§3. Presentation of the general combinations.

Convention 1. An expression X of the form
$$(\Re I Y_1 Y_2 \cdots Y_p x_1 x_2 \cdots x_n),$$

where the Y_i are entities, reduces *formally* to an expression Z if a reduction from X to Z takes place when the Y_i are treated as variables; or in case one wants to have it more precisely, when the expression $(\Re I x_1 x_2 \cdots x_{n+p})$ can be reduced to a Z' such that by putting Y_i in place of x_i for $i = 1, 2, \cdots, p$, and x_{i-p} in place of x_i for $i = p+1, p+2, p+n$ in Z', the expression Z is obtained.

SATZ 1. *Ist X ein Kombinator, so gibt es ein S der Form $(\mathfrak{R}IBCWK)$, das sich auf X formal reduziert.*

Beweis: Ersetzen wir in dem gegebenen Kombinator B, C, W, K durch x_1, x_2, x_3 bzw. x_4, so erzeugen wir eine Kombination Z von x_1, x_2, x_3, x_4. Nach §1, Sätzen 6 und 7 gibt es ein $\mathfrak{R}$, sodass $(\mathfrak{R}Ix_1x_2x_3x_4)$ sich auf Z reduziert. Daher reduziert $(\mathfrak{R}IBCWK)$ sich formal auf X, w.z.b.w.

SATZ 2. *X sei eine Kombination von Kombinatoren und Variablen $x_1, x_2, \cdots, x_n$, sodass*

a) *X auf einen ähnlichen X' durch einen einzigen Reduktionsprozess reduziert wird,*

b) *es ein S gibt, näm.*

(A) $$S \equiv \mathfrak{R}IY_1Y_2\cdots Y_p,$$

wo jedes Y_i entwerder B, C, W oder K ist, sodass der Ausdruck $(Sx_1x_2\cdots x_n)$ sich auf X formal reduziert.

Dann gibt es ein S', nämlich

(B) $$S' \equiv \mathfrak{R}'IY_1'Y_2'\cdots Y_q'$$

wo jedes Y_i^j entweder B, C, W oder K ist, sodass

α) $\vdash S' = S$,

β) *der Ausdruck $(S'x_1x_2\cdots x_n)$ sich auf X' formal reduziert.*

Beweis: $Y_0', Y_1', Y_2', \cdots, Y_q'$ seien die sämtlichen in X vorkommenden Grundkombinatoren (B, C, W oder K), und zwar so, dass jeder der Kombinatoren B, C, W, K unter diesen Y_i' genau so oft erscheint, wie in X selbst. Die Anordnung dieser Kombinatoren unter den Y_i' bleibt für jetzt gleichgültig. Die Y_i' kommen natürlich–abgesehen von ihrer Häufigkeit–unter den $Y_1, Y_2, \cdots, Y_p$ vor.

Behandeln wir nunmehr die $Y_1, Y_2, \cdots, Y_p$ formal als Variable, so schliessen wir die Folgenden:

1) Die Folge $IY_0'Y_1'\cdot Y_q'x_1x_2x_3\cdots$ ist eine Umwandlung der Folge $IY_1Y_2\cdots Y_px_1x_2x_3\cdots$,

2) Wenn wir die durch X bekstimmte Folge wie folgt schreiben, $y_0y_1y_2y_3\cdots$, wo y_0 entweder ein Y_i oder eine Variable ist, und y_i, $i > 0$, eine Kombination von $Y_1, Y_2, \cdots Y_p$ und Variablen ist, so ist die Folge

(1) $\quad\quad Iy_0y_1y_2\cdots$

Theorem 1. *If X is a combinator, then there is an S of the form $(\Re IBCWK)$ which reduces formally to X.*

Proof. If in the given combinator we replace B, C, W, K by x_1, x_2, x_3, resp. x_4, then we obtain a combination Z of x_1, x_2, x_3, x_4. By §1, Theorems 6 and 7 there is an $\Re$ such that $(\Re I x_1 x_2 x_3 x_4)$ reduces to Z. Hence $(\Re IBCWK)$ formally reduces to X, qed.

Theorem 2. *Let X be a combination of combinators and variables $x_1, x_2, \cdots, x_n$ such that*

a) *X reduces by a single reduction process to a similar X',*

b) *there is an S, namely*

(A) $$S \equiv \Re I Y_1 Y_2 \cdots Y_p,$$

where every Y_i is either B, C, W or K, such that the expression $(Sx_1 x_2 \cdots x_n)$ reduces formally to X.

Then there is an S', namely

(B) $$S' \equiv \Re' I Y_1' Y_2' \cdots Y_q'$$

where every Y_i' is either B, C, W or K, such that

α) $\vdash S' = S$,

β) *the expression $(S' x_1 x_2 \cdots x_n)$ reduces formally to X'.*

Proof. Let $Y_0', Y_1', Y_2', \cdots, Y_q'$ be all the basic combinators (B, C, W or K) appearing in X, and such that each of the combinators B, C, W, K appears inside these Y_i' exactly as often as in X itself. The arrangement of these combinators inside the Y_i' does not matter for now. Naturally the Y_i' – apart from their multiplicity – occur inside $Y_1, Y_2, \cdots, Y_p$.

If we now treat the $Y_1, Y_2, \cdots, Y_p$ formally as variables, then we will conclude the following:

1) The sequence $I Y_0' Y_1' \cdots Y_q' x_1 x_2 x_3 \cdots$ is a transformation of the sequence $I Y_1 Y_2 \cdots Y_p x_1 x_2 x_3 \cdots$,

2) If we write the sequence determined by X as $y_0 y_1 y_2 y_3 \cdots$, where y_0 is either a Y_i or a variable, and y_i, $i > 0$, is a combination of $Y_1, Y_2, \cdots Y_p$ and variables, then the sequence

(1) $\quad I y_0 y_1 y_2 \cdots$

das Produkt der eben erwähnten Umwandlung und eine Folge derselben Form wie (1).

Nun bezeichne ich mit Ω bzw. $\mathfrak{R}_1$ zwei normale Kombinatoren, sodass Ω bzw. $\mathfrak{R}_1 I$ Umwandlung bzw. der zuletzt erwähnten Folge entsprechen. Dann bemerken wir: 1) $(\Omega \cdot \mathfrak{R}_1)$ entspricht der Folge (1) (II C2, Satz 2); 2) wir dürfen annehmen dass $\mathfrak{R}_1$ und also $(\Omega \cdot \mathfrak{R}_1)^*$ kein Glied der Form B_n enthält (weil $\mathfrak{R}_1$ normal ist und $(\mathfrak{R}_1 I)$ einer Folge der Form (1) entspricht–vgl. Beweis von §1, Satz 6); 3) wir dürfen ferner annehmen, dass $\mathfrak{R}$ kein Glied der Form B_n hat (denn wenn $\vdash \mathfrak{R} = \mathfrak{R}^* \cdot B_n, \mathfrak{R}^*$ normal, so können wir in den Satz $\mathfrak{R}$ durch $\mathfrak{R}^*$ ersetzen). Daraus folgt

$$\vdash \mathfrak{R} = \Omega \cdot \mathfrak{R}_1 \qquad (\S1, \text{Satz 6}).$$
$$\therefore \vdash S = (\Omega \cdot \mathfrak{R}_1) I Y_1 Y_2 \cdots Y_p$$
$$= \mathfrak{R}_1 I Y_0' Y_1' \cdots Y_q' \qquad (\text{nach der Bedeutung von } \Omega).$$

Weiterhin reduziert der Ausdruck

(2) $\qquad (\mathfrak{R}_1 I Y_0' Y_1' \cdots Y_q' x_1 x_2 \cdots x_n)$

sich formal auf X. (Nach der Bedeutung von $\mathfrak{R}_1$, §1, Satz 7) .

Wir unterscheiden nun zwei Fälle; näm.–

I. Die Reduktion von X auf X' volzieht sich in dem ersten Sinne.

II. Die Reduktion von X auf X' vollzieht sich in dem zweiten Sinne.

Fall I. Hier sei Y_0' der erste in X vorkommende Grundkombinator. Dann erscheint Y_0' in X nur an der ersten Stelle. Deshalb ist X eine normale Kombination von $Y_0', Y_1', Y_2', \cdots, Y_q'$ und Variablen. Es gibt also, nach §1, Satz 5, ein $\mathfrak{R}_2$ und ein $\mathfrak{B}$, sodass

$$\vdash \mathfrak{R}_1 = B\mathfrak{R}_2 \cdot \mathfrak{B}.$$
$$\therefore \vdash \mathfrak{R}_1 I Y_0' = (\mathfrak{R}_2 \cdot \mathfrak{B} I) Y_0'$$
(3) $\qquad = (\mathfrak{R}_2 \cdot \mathfrak{B} I \cdot B Y_0') I \qquad (\S1, \text{Satz 3}).$

Nun betrachten wir Y_0' wieder als Kombinator und definieren:

a) $\mathfrak{R}' \equiv$ eine normale Form von $(\mathfrak{R}_2 \cdot \mathfrak{B} I \cdot B Y_0')$† ohne Glieder dor Form B_n,

b) $S' \equiv \mathfrak{R}' I Y_1' Y_2' \cdots Y_q'$,

so folgt $\vdash S' = S$. Also ist die Bedingung α) erfüllt.

Dieses $\mathfrak{R}' I$ entspricht, wenn wir $Y_1', Y_2', \cdots, Y_q'$ formal betrachten, der durch X' bestimmten Folge. Denn ich habe gezeigt, dass der Ausdruck (2)

---

*Sogar wenn es auf die Normalform gebracht wird.

†Dies ist regulär nach §1, Sätzen 1 und 2.

is the product of the transformation just mentioned and a sequence of the same form as (1).

Now denote with Ω and $\mathfrak{R}_1$ two normal combinators, such that Ω and $\mathfrak{R}_1 I$ conform, respectively, to this transformation and to the last mentioned sequence. Then we see that: 1) $(\Omega \cdot \mathfrak{R}_1)$ conforms to the sequence (1) (II C2, Theorem 2); 2) We can assume that $\mathfrak{R}_1$ and hence also $(\Omega \cdot \mathfrak{R}_1)^*$ contain no limbs of the form B_n (because $\mathfrak{R}_1$ is normal and $(\mathfrak{R}_1 I)$ conforms to a sequence of the form (1) –cf. the proof of §1, Theorem 6); 3) we can further assume that $\mathfrak{R}$ has no limbs of the form B_n (because if $\vdash \mathfrak{R} = \mathfrak{R}^* \cdot B_n$, $\mathfrak{R}^*$ is normal, then we can replace $\mathfrak{R}$ by $\mathfrak{R}^*$ in the theorem). Hence it follows that

$$\vdash \mathfrak{R} = \Omega \cdot \mathfrak{R}_1 \qquad (\S 1, \text{Thm. 6}).$$
$$\therefore \vdash S = (\Omega \cdot \mathfrak{R}_1) I Y_1 Y_2 \cdots Y_p$$
$$= \mathfrak{R}_1 I Y_0' Y_1' \cdots Y_q' \qquad (\text{by the meaning of } \Omega).$$

Furthermore the expression

(2) $\qquad (\mathfrak{R}_1 I Y_0' Y_1' \cdots Y_q' x_1 x_2 \cdots x_n)$

reduces formally to X. (By the meaning of $\mathfrak{R}_1$, §1, Theorem 7).

We distinguish now two cases; namely

I. The reduction from X to X' takes place in the first sense.

II. The reduction from X to X' takes place in the second sense.

Case I. Here let Y_0' be the first basic combinator that occurs in X. Then Y_0' only appears in X in the first argument. Hence X is a normal combination of $Y_0', Y_1', Y_2', \cdots, Y_q'$ and variables. Therefore there exist, by §1, Theorem 5, an $\mathfrak{R}_2$ and a $\mathfrak{B}$, such that

$$\vdash \mathfrak{R}_1 = B\mathfrak{R}_2 \cdot \mathfrak{B}.$$
$$\therefore \vdash \mathfrak{R}_1 I Y_0' = (\mathfrak{R}_2 \cdot \mathfrak{B} I) Y_0'$$
(3) $\qquad\qquad\qquad = (\mathfrak{R}_2 \cdot \mathfrak{B} I \cdot B Y_0') I \qquad (\S 1, \text{Thm. 3}).$

Now we consider Y_0' again as a combinator and define:

a) $\mathfrak{R}' \equiv$ a normal form of $(\mathfrak{R}_2 \cdot \mathfrak{B} I \cdot B Y_0')^\dagger$ without limbs of the form B_n,

b) $S' \equiv \mathfrak{R}' I Y_1' Y_2' \cdots Y_q'$,

hence $\vdash S' = S$ holds. Therefore condition $\alpha)$ is satisfied.

This $\mathfrak{R}' I$ conforms to the sequence determined by X' when we view $Y_1', Y_2', \cdots, Y_q'$ formally. For I have proved that the expression (2)

---

*Even if it is brought to a normal form.

†This is regular by §1, Theorems 1 and 2.

sich formal auf X reduziert. In dieser Reduktion betrachten wir Y_0' nunmehr nicht als Variable, sondern als Kombinator; dabei wird nichts in der Reduktion geändert. Die Reduktion lässt sich doch eine Stufe weiter auf X' durchführen (nach Hp. a). Aber weil

$$\vdash \mathfrak{R} = \mathfrak{R}_1 I Y_0' \qquad \text{(aus (3))},$$

und die beiden Seiten dieser Gleichung Folgen lauter Variablen entsprechen, so entsprechen sie derselben Folge (II C 1, Satz 11).

Dass die Bedingung β) erfüllt ist, folgt daraus nach §1, Satz 7.

Fall II. Hier soll Y_0' den Kombinator bezeichnen, welcher durch die Reduktion von X auf X' eliminiert wird. Er nehme in X die $(r+1)$te Stelle ein, wo $r > 0$ nach der Voraussetzung dieses Falles ist.

Nach II D6, Satz 1, gibt es $\mathfrak{R}_1, \mathfrak{W}_1, \mathfrak{C}_1$ und $\mathfrak{B}_1$ derart, dass

(4) $\qquad \vdash \mathfrak{R}_1 = \mathfrak{K}_1 \cdot \mathfrak{W}_1 \cdot \mathfrak{C}_1 \cdot \mathfrak{B}_1.$

Aber nach der Voraussetznng über $Y_0', Y_1', \cdots, Y_q'$ kann $\mathfrak{K}_1$ kein K_i für $i \leq q+1$ und $\mathfrak{W}_1$ kein W_j für $j \leq q+1$ enthalten, also wird

(5) $\qquad \vdash \mathfrak{K}_1 \cdot \mathfrak{W}_1 = B_{q+1}(\mathfrak{K}_2 \cdot \mathfrak{W}_2).$

Auch entspricht $\mathfrak{C}_1$ einer Permutationsfolge, welche in zwei Faktoren zerlegt werden kann, wie folgt: der erste Faktor lässt Y_0' invariant, aber ordnet $Y_1', Y_2', \cdots, Y_q'$ und die Variablen in die Anordnung, die sie in X haben, an; der zweite Faktor setzt Y_0' an die Stelle, die es in X hatt, aber lässt die Anordnung von $Y_1', \cdots, Yq'$ und die Variablen unter sich selbst, unverändert bleiben. Dem ersten Faktor entspricht ein $\mathfrak{C}$, dessen Glieder alle C_i mit $i > 1$ sind, also ein $\mathfrak{C}$ von der Form $B\mathfrak{C}_2$; dem zweiten Faktor entspricht Γ_r $(r > 0)$. Also (II D 5, Satz 7).

(6) $\qquad \vdash \mathfrak{C}_1 = B\mathfrak{C}_2 \cdot \Gamma_r.$

Daher (aus (4) (5) (6))

(7) $\qquad \vdash \mathfrak{R}_1 = B(B_q \mathfrak{K}_2 \cdot B_q \mathfrak{W}_2 \cdot \mathfrak{C}_2) \cdot \Gamma_r \cdot \mathfrak{B}.$

Nun erscheint Y_0' nach Hp. (a) und Definition am Anfang eines in X eingeklammerten Teilausdrucks; die Anzahl der Glieder ausser Y_0' dieses Teilausdrucks sei $m+1$. Dann (vgl. den Beweis von II C3, Satz 3, und II D 3, Satz 1) gibt es $\mathfrak{B}_2$ und $\mathfrak{B}_3$ derart, dass

(8) $\qquad \vdash \mathfrak{B} = B_{r+1}\mathfrak{B}_2 \cdot B_r B_{m+1} \cdot \mathfrak{B}_3.$

Weil die Glieder von Γ_r alle $C_1, C_2, \cdots$ oder C_r sind, so kann $B_{r+1}\mathfrak{B}_2$ mit allen diesen Gliedern, also mit Γ_r selbst, vertauscht werden (II D 2, Satz 5a).

reduces formally to X. In this reduction we now view Y_0' not as a variable, but as a combinator; hence nothing will change in the reduction. This reduction can be continued a stage further to X' (by Hyp. a). But because

$$\mathfrak{R}' = \mathfrak{R}_1 I Y_0' \quad \text{(by (3))},$$

and both sides of this equation conform to sequences of variables only, they conform to the same sequence (II C 1, Theorem 11).

That condition β) holds follows from above by §1, Theorem 7.

Case II. Here let Y_0' denote the combinator which is eliminated by the reduction from X to X'. Let it occupy in X the $(r+1)$th position, where $r > 0$ by the hypothesis of this case.

By II D 6, Theorem 1, there exist $\mathfrak{K}_1, \mathfrak{W}_1, \mathfrak{C}_1$ and $\mathfrak{B}_1$ such that

(4) $\quad \vdash \mathfrak{R}_1 = \mathfrak{K}_1 \cdot \mathfrak{W}_1 \cdot \mathfrak{C}_1 \cdot \mathfrak{B}_1.$

But by the hypotheses about $Y_0', Y_1', \cdots, Y_q'$, $\mathfrak{K}_1$ cannot contain any K_i for $i \leq q+1$ and $\mathfrak{W}_1$ cannot contain any W_j for $j \leq q+1$; hence

(5) $\quad \vdash \mathfrak{K}_1 \cdot \mathfrak{W}_1 = B_{q+1}(\mathfrak{K}_2 \cdot \mathfrak{W}_2).$

Also $\mathfrak{C}_1$ conforms to a permutation sequence, which can be decomposed into two factors as follows: the first factor leaves Y_0' invariant but arranges $Y_1', Y_2', \cdots, Y_q'$ and the variables in the arrangement which they had in X; the second factor places Y_0' in the location which it had in X but keeps the arrangement of $Y_1', \cdots, Y_q'$ and the variables inside itself unchanged. The first factor conforms to a $\mathfrak{C}$ whose limbs are all the C_i with $i > 1$, therefore a $\mathfrak{C}$ of the form $B\mathfrak{C}_2$; the second factor conforms to Γ_r ($r > 0$). Hence (II D 5, Theorem 7):

(6) $\quad \vdash \mathfrak{C}_1 = B\mathfrak{C}_2 \cdot \Gamma_r.$

Therefore (by (4) (5) (6))

(7) $\quad \vdash \mathfrak{R}_1 = B(B_q \mathfrak{K}_2 \cdot B_q \mathfrak{W}_2 \cdot \mathfrak{C}_2) \cdot \Gamma_r \cdot \mathfrak{B}.$

Now by Hyp. (a) and definition, Y_0' occurs at the start of a subexpression of X enclosed in parentheses; let the number of limbs of this subexpression excluding Y_0' be $m+1$. Hence (cf. the proof of II C3, Theorem 3, and II D3, Theorem 1) there are $\mathfrak{B}_2$ and $\mathfrak{B}_3$ where

(8) $\quad \vdash \mathfrak{B} = B_{r+1}\mathfrak{B}_2 \cdot B_r B_{m+1} \cdot \mathfrak{B}_3.$

Because all the limbs of Γ_r are $C_1, C_2, \cdots$, or C_r, then $B_{r+1}\mathfrak{B}_2$ can be exchanged with all these limbs, and hence with Γ_r itself (II D2, Theorem 5a).

Daher
$$\begin{aligned}
\vdash \mathfrak{R}_1 &= B(B_q\mathfrak{K}_2 \cdot B_q\mathfrak{W} \cdot \mathfrak{C}_2) \cdot \Gamma_r \cdot B_{r+1}\mathfrak{B}_2 \cdot B_r B_{m+1} \cdot \mathfrak{B}_3, \\
&= B(B_q\mathfrak{K}_2 \cdot B_q\mathfrak{W}_2 \cdot \mathfrak{C}_2 \cdot B_r\mathfrak{B}_2) \cdot \Gamma_r \cdot B_r B_{m+1} \cdot \mathfrak{B}_3, \\
&= B\mathfrak{R}_2 \cdot \Gamma_r \cdot B_r B_{m+1} \cdot \mathfrak{B}_3
\end{aligned}$$
(9)

wenn ich nur definiere:
$$\mathfrak{R}_2 \equiv B_q\mathfrak{K}_2 \cdot B_q\mathfrak{W}_2 \cdot \mathfrak{C}_2 \cdot B_r\mathfrak{B}_2.$$

Daher
$$\begin{aligned}
\vdash \mathfrak{R}_1 I Y_0' &= (B\mathfrak{R}_2 \cdot \Gamma_r \cdot B_r B_{m+1} \cdot \mathfrak{B}_3) I Y_0' \\
&= B\mathfrak{R}_2(\Gamma_r((B_r B_{m+1} \cdot \mathfrak{B}_3)I))Y_0' \\
&= \mathfrak{R}_2(\Gamma_r(B_{r-1}B_{m+1} \cdot \mathfrak{B}_3 I)Y_0').
\end{aligned}$$
(10)

Weil nach Hp. a) und Definition von Y_0' eine Reduktion durch Y_0' wirklich stattfindet, so muss $m \geq 2$ sein, wenn $Y_0' B$ oder C ist, und $m \geq 1$, wenn $Y_0' W$ oder K ist. Infolgedessen muss es nach II D 2, Satz 2, und den kommutativen Axiomen ein n geben, wofür $\vdash C_1 B_{m+1} Y_0', = BY_0' \cdot B_n$. Also

$$\vdash \Gamma_r(B_{r-1}B_{m+1} \cdot \mathfrak{B}_3 I)Y_0' = B_r Y_0' \cdot B_{r-1} B_n \cdot (\mathfrak{B}_3 I) \quad (\S 2, \text{Satz } 7);$$

also, wenn wir dies in (10) einsetzen,

$$\begin{aligned}
\vdash \mathfrak{R}_1 I Y_0' &= \mathfrak{R}_2(B_r Y_0' \cdot B_{r-1} B_n \cdot \mathfrak{B}_3 I) \\
&= (\mathfrak{R}_2 \cdot B_{r+1} Y_0' \cdot B_r B_n \cdot \mathfrak{B}_3)I \quad (\S 1, \text{Satz } 3).
\end{aligned}$$
(11)

Definiere ich nun

a) $\qquad \mathfrak{R}' \equiv \mathfrak{R}_2 \cdot B_{r+1} Y_0' \cdot B_r B_n \cdot \mathfrak{B}_3,$

b) $\qquad S' \equiv \mathfrak{R}' I Y_1' Y_2' \cdots Y_q',$

so folgt aus (11) und (A), dass $\vdash S = S'$.

Dass die Bedingung β) erfüllt ist, folgt hier genau wie im Fall I.

SATZ 3. *Ist X ein solcher Kombinator, dass*

(1) $\qquad (X x_1 x_2 \cdots x_n)$

sich auf eine Kombination von $x_1, x_2, \cdots, x_n$ reduziert; dann lässt X sich in eine ($\mathfrak{R}I$) umformen und zwar so, dass ($\mathfrak{R}I x_1 x_2 \cdot x_n$) sich auf die gegebene Kombination reduziert.

Beweis: Nach den Voraussetzungen gibt es eine Reihe von Ausdrücken $X_1, X_2, \cdots, X_m$ derart, dass 1) X_{i+1} sich aus X_i durch einen einzigen Reduktionsprozess erzielt, 2) X_1 mit dem Ausdruck (1) identisch ist, 3) X_m eine Kombination von $x_1, x_2, \cdots, x_n$ ist.

Wir können nun diesen X_i eine Reihe von Kombinatoren $S_1, S_2, \cdots, S_m$ zuordnen und zwar so dass

a) Jedes S_i in der Form (A) (s. Satz 2) steht,

b) $(S_i x_1 x_2 \cdots x_n)$ sich auf X_i formal reduziert,

c) $\vdash S_{i+1} = S_i$.

Hence $\vdash \mathfrak{R}_1 = B(B_q\mathfrak{K}_2 \cdot B_q\mathfrak{W}_2 \cdot \mathfrak{C}_2) \cdot \Gamma_r \cdot B_{r+1}\mathfrak{B}_2 \cdot B_r B_{m+1} \cdot \mathfrak{B}_3,$
$= B(B_q\mathfrak{K}_2 \cdot B_q\mathfrak{W}_2 \cdot \mathfrak{C}_2 \cdot B_r\mathfrak{B}_2) \cdot \Gamma_r \cdot B_r B_{m+1} \cdot \mathfrak{B}_3,$
(9) $= B\mathfrak{R}_2 \cdot \Gamma_r \cdot B_r B_{m+1} \cdot \mathfrak{B}_3$

if I define:
$$\mathfrak{R}_2 \equiv B_q\mathfrak{K}_2 \cdot B_q\mathfrak{W}_2 \cdot \mathfrak{C}_2 \cdot B_r\mathfrak{B}_2.$$
Hence $\vdash \mathfrak{R}_1 I Y_0' = (B\mathfrak{R}_2 \cdot \Gamma_r \cdot B_r B_{m+1} \cdot \mathfrak{B}_3) I Y_0'$
$= B\mathfrak{R}_2(\Gamma_r((B_r B_{m+1} \cdot \mathfrak{B}_3)I))Y_0'$
(10) $= \mathfrak{R}_2(\Gamma_r(B_{r-1} B_{m+1} \cdot \mathfrak{B}_3 I)Y_0').$

Because by Hyp. a) and the definition of Y_0' a reduction involving Y_0' really takes place, then it must be the case that $m \geqq 2$ if Y_0' is B or C, and $m \geqq 1$ if Y_0' is W or K. Consequently, by II D2, Theorem 2, and the commutative axioms there must exist an n for which $\vdash C_1 B_{m+1} Y_0' = BY_0' \cdot B_n$. Hence

$\vdash \Gamma_r(B_{r-1} B_{m+1} \cdot \mathfrak{B}_3 I) Y_0' = B_r Y_0' \cdot B_{r-1} B_n \cdot (\mathfrak{B}_3 I)$ (§2, Theorem 7);

thus when we put this in (10),

$\vdash \mathfrak{R}_1 I Y_0' = \mathfrak{R}_2(B_r Y_0' \cdot B_{r-1} B_n \cdot \mathfrak{B}_3 I)$
(11) $= (\mathfrak{R}_2 \cdot B_{r+1} Y_0' \cdot B_r B_n \cdot \mathfrak{B}_3) I$ (§1, Theorem 3).

Now I define

a) $\mathfrak{R}' \equiv \mathfrak{R}_2 \cdot B_{r+1} Y_0' \cdot B_r B_n \cdot \mathfrak{B}_3,$
b) $S' \equiv \mathfrak{R}' I Y_1' Y_2' \cdots Y_q',$

hence by (11) and (A) it follows that $\vdash S = S'$.

That condition β) is satisfied follows here exactly as in Case I.

Theorem 3. *If X is a combinator such that*

(1) $(X x_1 x_2 \cdots x_n)$

reduces to a combination of $x_1, x_2, \cdots x_n$; then X can be converted into an $(\mathfrak{R} I)$ such that $(\mathfrak{R} I x_1 x_2 \cdots x_n)$ reduces to the given combination.

Proof. By the hypotheses there is a series of expressions $X_1, X_2, \cdots, X_m$ such that 1) X_{i+1} is reached from X_i by a single reduction process, 2) X_1 is identical to the expression (1), (3) X_m is a combination of $x_1, x_2, \cdots, x_n$.

We can now associate to these X_i as a series of combinators $S_1, S_2, \cdots, S_m$ such that

a) Every S_i is in the form (A) (see Theorem 2),
b) $(S_i x_1 x_2 \cdots x_n)$ can be formally reduced to X_i,
c) $\vdash S_{i+1} = S_i$.

In der Tat gilt als S_1 der in Satz 1 ausgestellte Kombinator; und aus Satz 2 folgt, dass aus einem gegebenen S_i, $(i < m)$ ein S_{i+1} konstruiert werden kann.

In dieser Weise haben wir ein S_m, etwa

(2) $\quad S_m \equiv \mathfrak{R}_m I Y_1 Y_2 \cdots Y_p \quad (Y_i \equiv B, C, W \text{ oder } K, \mathfrak{R}_m \text{ normal})$

sodass $(S_m x_1 x_2 \cdots x_n)$ sich *formal* auf eine Kombination lauter Variablen reduziert. In dieser Reduktion müssen freilich alle $Y_1, Y_2, \cdots, Y_p$ ausfallen. Also wenn $\mathfrak{R}_m$ auf die normale Form gebracht wird, gilt

$$\vdash \mathfrak{R}_m = K_p \cdot K_{p-1} \cdots K_1 \cdot \mathfrak{R}'_m.$$

Infolgedessen	$\vdash S_m$	$= \mathfrak{R}'_m I.$	(aus (2), II B 3).
Aber	$\vdash S_m$	$= S_1$	(aus c)),
		$= X$	(Bedeutung von S_1).
$\therefore$	$\vdash X$	$= \mathfrak{R}'_m I.$	w.z.b.w.

SATZ 4. *Wenn zwei Kombinatoren Y_1 und Y_2 derselben Folge lauter Variablen entsprechen*;
dann $\quad\quad\quad \vdash Y_1 = Y_2$.

Beweis: Nach Satz 3 gibt es $\mathfrak{R}_1$ und $\mathfrak{R}_2$, sodass

$$\vdash Y_1 = \mathfrak{R}_1 I \quad\quad \vdash Y_2 = \mathfrak{R}_2 I,$$

und die beiden Kombinatoren $(\mathfrak{R}_1 I)$ und $(\mathfrak{R}_2 I)$ auch derselben Folge entsprechen. Wir können ohne Beschränkung der Allgemeinheit annehmen, dass $\mathfrak{R}_1$ und $\mathfrak{R}_2$ normal und ohne Glieder der Form B_n sind.

Dann	$\vdash \mathfrak{R}_1$	$= \mathfrak{R}_2$	(§1, Satz 6).
Also	$\vdash Y_1$	$= Y_2$.	w.z.b.w.

SATZ 5. *Wenn zwei eigentliche Kombinatoren Y_1 und Y_2 dieselbe Kombination von lauter Variablen darstellen, dann* $\vdash Y_1 = Y_2$.

Beweis: Klar aus Satz 4.

§4. *Die Substitutionsprozesse.*

Zum Schluss gebe ich hier einige Sätze über die Verhältnisse der Substitutionsprozesse zu den Kombinatoren. Die Bewiese gebe ich nur kurz, weil sie meistens nur Bechnungsübungen sind.

Die Substitutionsprozesse lassen sich zunächst durch Kombinationen von Variablen darstellen. Z. B. betrachten wir den Ausdruck:

$$(ux_1(vx_2x_3)x_4).$$

Wenn u und v Grundfunktionen sind, so bedeutet dies eine gewisse aus einer

In fact the combinator constructed in Theorem 1 serves as S_1; and by Theorem 2 it follows from a given S_i ($i < m$), an S_{i+1} can be constructed.

In this fashion we have an S_m, like

(2) $\quad S_m \equiv \mathfrak{R}_m I Y_1 Y_2 \cdots Y_p \quad (Y_i \equiv B, C, W \text{ or } K, \mathfrak{R}_m \text{ normal})$

such that $(S_m x_1 x_2 \cdots x_n)$ reduces formally to a combination of variables only. In this reduction certainly all of $Y_i, Y_2, \cdots Y_p$ must disappear. Hence if $\mathfrak{R}_m$ is brought to a normal form, we have

$$\vdash \mathfrak{R}_m = K_p \cdot K_{p-1} \cdots K_1 \cdot \mathfrak{R}'_m.$$

Consequently $\vdash S_m = \mathfrak{R}'_m I.$ (by (2), II B 3).
But $\qquad \vdash S_m = S_1$ (by c)),
$\qquad\qquad\;\; = X$ (meaning of S_1).
$\therefore \qquad\quad \vdash X = \mathfrak{R}'_m I.$ qed.

Theorem 4. *If two combinators Y_1 and Y_2 conform to the same sequence of variables only;*
then $\qquad\qquad \vdash Y_1 = Y_2.$

Proof. By Theorem 3 there exist $\mathfrak{R}_1$ and $\mathfrak{R}_2$, such that

$$\vdash Y_1 = \mathfrak{R}_1 I \qquad \vdash Y_2 = \mathfrak{R}_2 I,$$

and the two combinators $(\mathfrak{R}_1 I)$ and $(\mathfrak{R}_2 I)$ also conform to the same sequence. Without loss of generality we can assume that $\mathfrak{R}_1$ and $\mathfrak{R}_2$ are normal and without limbs of the form B_n.

Then $\qquad\qquad \vdash \mathfrak{R}_1 = \mathfrak{R}_2$ (§1, Theorem 6).
Thus $\qquad\qquad \vdash Y_1 = Y_2$ qed.

Theorem 5. *If two pure combinators Y_1 and Y_2 represent the same combination of variables only, then $\vdash Y_1 = Y_2$.*

Proof. Obvious by Theorem 4.

§4. *Substitution processes.*

To conclude I give here some theorems about the relation of substitution processes to the combinators. The proofs I only give in outline because they are mostly only exercises in computation.

The substitution processes can be initially represented by combinations of variables. As an example we consider the expression:

$$(ux_1(vx_2x_3)x_4).$$

If u and v are basic functions, then this means the function of x_1, x_2, x_3, x_4 generated from a certain mode

Verknüpfung von u und v erzeugte Funktion von x_1, x_2, x_3, x_4. Aber wir können ihn auch,–wenn wir u und v für Variablen halten–als eine Funktion von u und v betrachten, welche für bestimmte Werte von u und v jene Funktion von x_1, x_2, x_3, x_4 darstellt–d. h. als den Verknüpfungsprozess selbst. Diese Auffassung ist naturgemäss, weil nach der Ausdeutung von Anwendung der Ausdruck für irgendeine bestimmten Werte von u, v, x_1, x_2, x_3, x_4 die mit der Auffassung verträgliche Aussage bedeutet. Der Ausdruck lässt sich ferner in

$$(C_1 \cdot BB_2)uvx_1x_2x_3x_4$$

umformen. Von unserem Gesichtspunkte aus ist also $(C_1 \cdot BB_2)$ der Substitutionsprozess selbst–eine Funktion, welche aus u und v die Funktion $((C_1 \cdot BB_2)uv)$ liefert, wo diese letzte die Funktion ist, welche aus x_1, x_2, x_3 die eben geschilderte Aussage liefert. In diesem Sinne können wir sagen, dass $(C_1 \cdot BB_2)$ den betreffenden Substitutionsprozess *darstellt*.

Von diesem Gesichtspunkt aus haben wir die folgenden Satze:

SATZ 1. *Jede Umwandlung im Sinne von Abschnitt A lässt sich durch ein Ω darstellen.*

SATZ 2. *Die Einsetzung von einer Funktion als Funktion von n Variablen an die Stelle der $(m+1)$ ten Variablen einer zweiten wird durch $(\Gamma_m \cdot B_m B_n)$ dargestellt.*

SATZ 3. *Sind die Substitutionsprozesse wie in den Sätzen 1 und 2 dargetellt, dann gestalten sie Ausdrücke der Form $(Yu_1u_2 \cdots u_n)$, wo Y eine eigentliche Kombination von Ordnung nicht zu gross ist, in andere Ausdrücke derselben Form um.*

Beweis: Für eine Umwandlung gilt

$$\begin{aligned}\vdash \Omega(Yu_1u_2 \cdots u_n) &= B_n\Omega Yu_1u_2 \cdots u_n \\ &= (B_{n-1}\Omega \cdot Y)u_1u_2 \cdots u_n.\end{aligned}$$

Für Zusammensetzungen: es sei $Z \equiv \Gamma_p \cdot B_pB_q$; dann

$$\begin{aligned}\vdash Z(Xu_1u_2 \cdots u_m)(Yv_1v_2 \cdots v_n) \\ = (B_mZ \cdot BX)Iu_1u_2 \cdots u_m(Yv_1v_2 \cdots v_n) \\ = (\Gamma_m \cdot B_mB_n \cdot B_mZ \cdot BX)IYu_1u_2 \cdot u_mv_1v_2 \cdots v_n \\ = Uu_1u_2 \cdots u_mv_1v_2 \cdots v_m,\end{aligned}$$

wo U eigentlich ist, wenn nur X und Y eigentlich sind, und q gross genug ist, sodass $Yx_1x_2 \cdots x_{n+q}$ sich auf eine Kombination lauter Variablen reduziert. In der Tat reduziert $(Uu_1u_2 \cdots u_mv_1v_2 \cdots v_nx_1x_2 \cdots x_{p+q})$ sich

of combination of u and v. But we can also consider it as a function of u and v –when think of u and v as variables– which for given values of u and v represents that function of x_1, x_2, x_3, x_4 –i.e. as the combination process itself. This interpretation is natural because, by the meaning of application, the expression denotes for any particular values of u, v, x_1, x_2, x_3, x_4 a statement that is compatible with this interpretation. The expression can be further transformed into

$$(C_1 \cdot BB_2)uvx_1x_2x_3x_4.$$

From our point of view $(C_1 \cdot BB_2)$ is therefore the substitution process itself – a function, which delivers for u and v the function $((C_1 \cdot BB_2)uv)$, where this latter is the function which from x_1, x_2, x_3 delivers the function just described. In this sense we say that $(C_1 \cdot BB_2)$ *represents* the substitution process in question.

From this point of view we have the following theorems:

Theorem 1. *Every transformation in the sense of Section A can be represented by an Ω.*

Theorem 2. *The replacement, by a function, as a function of n variables, of the $(m+1)$th variable in a second function is represented by $(\Gamma_m \cdot B_m B_n)$.*

Theorem 3. *If the substitution processes are represented as in Theorems 1 and 2, then they transform expressions of the form $(Yu_1u_2\cdots u_n)$, where Y is a proper combination of not too large an order, into other expressions of the same form.*

Proof. For a transformation we have

$$\vdash \Omega(Yu_1u_2\cdots u_n) = B_n\Omega Yu_1u_2\cdots u_n$$
$$= (B_{n-1}\Omega \cdot Y)u_1u_2\cdots u_n.$$

For a replacement: let $Z \equiv \Gamma_p \cdot B_p B_q$; then

$$\vdash Z(Xu_1u_2\cdots u_m)(Yv_1v_2\cdots v_n)$$
$$= (B_m Z \cdot BX)Iu_1u_2\cdots u_m(Yv_1v_2\cdots v_n)$$
$$= (\Gamma_m \cdot B_m B_n \cdot B_m Z \cdot BX)IYu_1u_2\cdots u_m v_1v_2\cdots v_m$$
$$= Uu_1u_2\cdots u_m v_1v_2\cdots v_n,$$

where U is proper, if X and Y are proper, and q is big enough so that $Yx_1x_2\cdots x_{n+q}$ reduces to

auf $(Xu_1u_2\cdots u_mx_1x_2\cdots x_p(Yv_1v_2\cdots v_nx_{p+1}\cdot x_{p+q}))$. Die Bedingung auf Y ist erfüllt, wenn wir es mit einem Substitutionsprozess zu tun haben.

SATZ 4. *X sei eine Kombination von Variablen und gewissen Etwasen* $u_1, u_2, \cdots, u_m$ *Dann gibt es einen Kombinator Y, sodass*

$$Yu_1u_2u_mx_1x_2\cdots x_n = X, \quad (u_i \text{ als Variable behandelt}).$$

Gibt es weiter einen Kombinator Z, sodass

$$\vdash Z_1v_2v\cdots v_px_1x_2\cdots x_n = X \quad (v_i \text{ als Variable betrachtet}),$$
wo $\quad \vdash Iv_1v_2\cdots v_p = VIu_1u_2\cdots u_m, \quad$ (V ein Kombinator),
(oder umgekehrt),
dann $\quad \vdash Yu_1u_2\cdots u_m = Zv_1v_2\cdots v_p.$

Beweis:* Wenn die $v_1v_2\cdots v_p$ dieselbe Reihe von Etwasen bildet, wie $u_1u_2,\cdots,u_m$, so folgt der Satz aus §3, Satz 5. Sonst

$$\vdash Zv_1v_2\cdots v_p = (V\cdot BZ)Iu_1u_2\cdots u_p,$$
und $\quad\quad\quad \vdash (V\cdot BZ)I = Y \quad\quad\quad$ (§3, Satz 5).
$\therefore \quad\quad \vdash Zv_1v_2\cdots v_p = Yu_1u_2\cdots u_m \quad\quad$ w.z.b.w.

---

*Der Beweis des ersten Teils des Satzes ist klar (§1,Satz 8).

a combination of variables only. In fact

$$(Uu_1u_2\cdots u_mv_1v_2\cdots v_nx_1x_2\cdots x_{p+q})$$

reduces to

$$(Xu_1u_2\cdots u_mx_1x_2\cdots x_p(Yv_1v_2\cdots v_nx_{p+1}\cdots x_{p+q})).$$

The condition for Y is satisfied when we are dealing with a substitution process.

Theorem 4. *Let X be a combination of variables and some entities u_1, u_2, $\cdots$, u_m. Then there is a proper combinator Y such that*

$$Yu_1u_2\cdots u_mx_1x_2\cdots x_n = X, \quad (u_i \text{ treated as variable}).$$

Furthermore, if there is a proper combinator Z such that

$\vdash Zv_1v_2\cdots v_px_1x_2\cdots x_n = X$ (v_i *treated as variable*),
where $\vdash Iv_1v_2\cdots v_p = VIu_1u_2\cdots u_m,$ (V *a combinator*),
(or otherwise),
then $\vdash Yu_1u_2\cdots u_m = Zv_1v_2\cdots v_p.$

Proof.* If the $v_1v_2\cdots v_p$ builds the same series of entities as $u_1, u_2, \cdots, u_m$, then the theorem follows by §3, Theorem 5. Otherwise

$$\begin{aligned} \vdash Zv_1v_2\cdots v_p &= (V\cdot BZ)Iu_1u_2\cdots u_p, \\ \text{and} \quad \vdash (V\cdot BZ)I &= Y \quad\quad (\S3, \text{ Theorem 5}). \\ \therefore \quad \vdash Zv_1v_2\cdots v_p &= Yu_1u_2\cdots u_m \quad\quad \text{qed.} \end{aligned}$$

---

*The proof of the first part of the theorem is obvious (§1, Theorem 8).

Notes

[1] Curry coined the word "contensive" in English to translate *"inhaltlich"* which expresses the idea of meaningfulness that is independent of any meaning given by the definition of the theory.

[2] This is a reference to the rule of detachment in *Principia Mathematica*, which is essentially the rule of *modus ponens*.

[3] Hereafter the word "blank" will be used instead of argument place.

[4] This is a description of the process now known as *currying*, a process named for Curry. Curry himself, once he became aware of this term, protested that he got the idea from Schönfinkel, so it should not be named for him (although this so called currying process first appeared in 1879 in Frege's Begriffschrift [32]). But can anybody imagine talking about "schönfinkeling" a function?

[5] This is the way the original text reads, but since Y could contain occurrences of "numbered free variables," those free variables should be counted among the $u_1, u_2, \cdots, u_m$.

[6] Curry's German is "Die Kategorie Begriff."

[7] Curry used the word "entity" in his early papers in English as a noun denoting into English of the German word *Etwas*. Then, after he wrote [14], he was informed by a philosopher that his use of this word involved philosophical conclusions that he did not wish to make. See [15, p. 157]. He therefore changed to the word "term," which he used in several papers. But for a formal system for quantification logic, where the word "term" referred to what we usually call "formulas" and there are other formal objects which are usually called "terms" he became dissatisfied with this usage. Curry always wanted a vocabulary that would not need exceptions to the rules. So he coined his own word: "ob," which is the first syllable of "object." From then on, he used this word "ob" for all formal objects with the property that each one has a unique construction from the atomic obs.

[8] It is usually called *modus ponens*.

[9] The notations $\equiv_{x,y}$ and $\equiv_x$ are from Russell and Whitehead *Principia Mathematica*, and express general equivalence for all possible values of the variable subscripts. In 1928 or 1929, when Curry wrote this, he could assume that most readers of his dissertation would be familiar with that notation.

[10] Thanks to Wilfried Buchholz for help in translating this expression.

[11] This is what the original text says. However, this is clearly not true for all combinators X, for example $X = CI$.

[12] This sentence is not completely clear, since it is not clear what X is for an arbitrary combinator Y. But this text does accurately translate the sentence in the German original.

[13] Here is the beginning of the second part.

[14] There are earlier references to and uses of variables. Here is where Curry starts explaining what he means by the term "variable".

[15] The reduction here is now called *weak reduction*, and does not correspond to either $\lambda\beta$- or $\lambda\beta\eta$-reduction.

[16] This is now called a *head contraction*. If it is repeated, it is called a *head reduction*.

[17] This is now called a *leftmost contraction* (but not a head contraction).

[18] A more literal translation of the sentence in the German original would be, "The highest variable appearing essentially in X is called the *grade* of X."

[19] This is implied by Part 2 of Convention 2 above; Curry is referring to reductions which are now called leftmost. This is clearly false as stated if we use the most general weak reductions allowed today, since $(Kx_1x_2)(Kx_3x_4)$ reduces to both $x_1(Kx_3x_4)$ and $(Kx_1x_2)x_3$, which do not reduce to each other.

[20] Cf. the n in the proof of Theorem 1.

[21] This is not the current use of the phrase "normal form."

[22] The German original reads here, *"Gibt es überdies in andreres der Folge entsprechendes $\mathfrak{R}_2$,"* which says "if there is another such $\mathfrak{R}_2$ which conforms to the sequence". We think this is a misprint in the original German text.

A Errata in Curry's thesis

- Page 24 (512 in Curry's original), line 2 from the bottom (last line in Curry's original), for "letzten" read "linkern". (This correction was made by Curry himself in one of his copies of the dissertation.)
- Page 26 (513 in Curry's original), For equation (2) read "$Y u_1 u_2 \cdots u_m$".
- Page 32 (516 in Curry's original), line 17 from bottom, for "Schöfinkel" read "Schönfinkel".
- Page 38 (519 in Curry's original), line 10, Curry uses "*bzw.*" here, but it appears that "*d.h.*," meaning "*i.e.*," would have served better.
- Page 42 (522 in Curry's original), line 19, for

$$\vdash BBW = B(B(B(B(BW)W)(BC))B(BB))B$$

read "$\vdash BBW = B(B(B(B(BW)W)(BC))(B(BB)))B$".
- Page 44 (522 in Curry's original), line 20, One parenthesis missing. Should read "$\vdash PX(PY(\Lambda XY))$".
- Page 46 (523 in Curry's original), line 4, For "$Q(CQXX)(CQXX)$" read "$Q(CQXX)(CQXY)$".
- Page 46 (523 in Curry's original), line -14, there is a missing parenthesis. I.e., the formula should read "$\vdash Q(C(B(WK))ZY)(YZ)$".
- Page 48 (524 in Curry's original), line -2, For "*bediesen*" read "*bewiesen*".
- Page 50 (525 in Curry's original), line 10, For "$X_{n=1}$" read "X_{n-1}".
- Page 60 (530 in Curry's original), line 9, For "(YZ)" read "(YZ_1)".
- Page 66 (533 in Curry's original), line 18, For "$X \cdot (Y \cdot Z)$" read "$(X \cdot Y) \cdot Z$".
- Page 68 (534 in Curry's original), line 9, For "$(m, n, n = 0, 1, 2, \cdots)$" read $(m, n, p = 0, 1, 2, \cdots)$".
- Page 70 (790, line 9 in Curry's original), line 10, for "betrefienden" read "betreffenden".
- Page 80 (792, line -10 in Curry's original), line -11, for "X aus Y" read "Y aus "X".
- Page 84 (794 in Curry's original), line 23, For the first "X''''" read "X'''''".
- Page 84 (794 in Curry's original), line 24, 'For '$(XX_3 \cdots X_p)$" read "$(X_1 X_3 \cdots X_p)$".
- Page 86 (795, line -6 in Curry's original), line -6, for $\mathfrak{A}_{n+1}$" read "$\mathfrak{A}_{N+1}$".
- Page 96 (800 in Curry's original), line 19, For "x_q" read "z_q".
- Page 110 (807 in Curry's original), line 18, For "γ" read "γ'".
- Page 112 (808 in Curry's original), line 7, For "$(\kappa' \cdot \omega' \cdot \pi) = (\kappa \cdot \omega \cdot \pi')$" read "$(\kappa' \cdot \omega' \cdot \pi') = (\kappa \cdot \omega \cdot \pi)$".

- Page 114 (809 in Curry's original), line -1, For "Satz 3" read "Satz 6".
- Page 116 (810 in Curry's original), line 5, For "Satz 2" read "Satz 1".
- Page 116 (810 in Curry's original), lines 16..20, For all "B_n" read "B_k".
- Page 116 (810 in Curry's original), line -3, For "X" read "Y".
- Page 118 (811 in Curry's original), line 8, For "X" read "Y".
- Page 118 (811 in Curry's original), line 10, For "$m \geqq p \geq 1$" read "$m > p \geqq 1$". The error was noted by Curry in a marginal note with a further note that the error was noted by Rosser in [48, p. 338]. Curry further noted in the margin, with a date of Jan 5, 1933, that the original manuscript was correct.
- Page 120 (812 in Curry's original), line 5, For "und §1, Satz 1" read "und Satz 1".
- Page 120 (812 in Curry's original), line 19, For "eben" read "oben".
- Page 122 (813 in Curry's original), line 10, For "$B_{m+1}B$" read "$B_{m-1}B$".
- Page 124 (814 in Curry's original), line 9, for "*Für jedes*" read "*Zu jedem*". (This correction was noted by Curry himself in a marginal note in one of his copies of the dissertation.)
- Page 126 (815 in Curry's original), line 2, For "W_{r+1}" read "W_{p+1}".
- Page 126 (815 in Curry's original), line 5, For "$W_{m+3} \cdot W_{m+2}$" read "W_{m+3}".
- Page 130 (817 in Curry's original), line 8, For "B_m" read "B_{m-1}".
- Page 130 (817 in Curry's original), line 9, For "B_{m+2}" read "B_{m+1}".
- Page 132 (818 in Curry's original), line -11, For "C_{j-1}" read "C_j".
- Page 136 (820 in Curry's original), line 5, For "§4" read "§5".
- Page 136 (820 in Curry's original), line -9, For "$\mathfrak{B}$" read "$\mathfrak{B}_2$".
- Page 144 (824 in Curry's original), line 20, After "*der Folge entsprechendes* $\mathfrak{R}_2$' insert (before the comma) "*sodas* $(\mathfrak{R}_2 I)$ *der Folge entspricht*".
- Page 146 (825 in Curry's original), line 6, For "$\mathfrak{R}_2 = \mathfrak{R}'_2 \cdot B_n$" read "$\mathfrak{R}'_2 = \mathfrak{R}_2 \cdot B_n$". 68
- Page 148 (826 in Curry's original), line 6, For "$X_1 Y$" read "$Y X_1$" and for "X_n" read "X_{n+1}".
- Page 148 (826 in Curry's original), line -1, For "$\equiv$" read "$=$".
- Page 150 (827 in Curry's original), line 16, For "*Wenn XY Etwase sind*" read "*Wenn X, Y Etwase sind*".
- Page 154 (829, line - 5 in Curry's original), line -7, for "for Form" read "der Form".
- Page 158 (831 in Curry's original), line 1, For "$\mathfrak{W}$" read "$\mathfrak{W}_2$".

- Page 162 (833 in Curry's original), line -5, For "$u_1u_2.u_m$" read "$u_1u_2\cdots u_m$".
- Page 164 (834 in Curry's original), lines 4 and 6, for "*einen Kombinator*" read "*einen eigentlichen Kombinator*".
- Page 164 (834 in Curry's original), line 5, for "$Yu_1u_2u_mx_1x_2\cdots x_n = X$" read "$Yu_1u_2\cdots u_mx_1x_2\cdots x_n = X$".
- Page 164 (834 in Curry's original), line 7, For "Z_1v_2v" read "Zv_1v_2".